Discovery Travel Adventures™

AMERICAN SAFARI

Judith Dunham
Editor

John Gattuso
Series Editor

D1372383

Discovery Communications, Inc.

Discovery Communications, Inc.

John S. Hendricks, *Founder, Chairman, and Chief Executive Officer*
Judith A. McHale, *President and Chief Operating Officer*
Michela English, *President, Discovery Enterprises Worldwide*

Discovery Publishing

Ray Cooper, *Vice President, Publishing*
Natalie Chapman, *Publishing Director*
Rita Thievon Mullin, *Editorial Director*
Mary Kalamaras, *Senior Editor*
Kimberly Small, *Senior Marketing Manager*
Chris Alvarez, *Business Development & Operations*
Tracy Fortini, *Discovery Channel Retail*
Steve Manning, *Naturalist, The Nature Company*

Insight Guides

Jeremy Westwood, *Managing Director*
Brian Bell, *Editorial Director*
John Gattuso, *Series Editor*
Siu-Li Low, *General Manager, Books*

Distribution

United States and Canada
Langenscheidt Publishers, Inc.
46-35 54th Road
Maspeth, NY 11378
Fax: 718-784-0640

Worldwide
APA Publications GmbH & Co.
Verlag KG Singapore Branch, Singapore
38 Joo Koon Road, Singapore 628990
Tel: 65-865-1600. Fax: 65-861-6438

Discovery Communications produces high-quality nonfiction television programming, interactive media, books, films, and consumer products. Discovery Networks, a division of Discovery Communications, Inc., operates and manages the Discovery Channel, TLC, Animal Planet, and Travel Channel. Visit Discovery Channel Online at http://www.discovery.com.

Although every effort is made to provide accurate information in this publication, we would appreciate readers calling our attention to any errors or outdated information by writing us at: Insight Guides, PO Box 7910, London SE1 1WE, England; fax: 44-171-403 0290; email: insight@apaguide.demon.co.uk

Printed by Insight Press Services (Pte) Ltd, 38 Joo Koon Road, Singapore 628990.

Library of Congress Cataloging-in-Publication Data
American Safari/Judith Dunham, editor; John Gattuso, series editor.
        p. cm.—(Discovery travel adventures)
    Includes bibliographical references.
    ISBN 1-56331-834-2 (pbk.)
    1. Wildlife viewing sites—United States Guidebooks.
2. Wildlife watching—United States Guidebooks. 3. Natural history—United States Guidebooks. 4. United States Guidebooks. I. Dunham, Judith. II. Series.
    QL 155.A555 1999
    508.73—dc21
                        99-24594
                        CIP

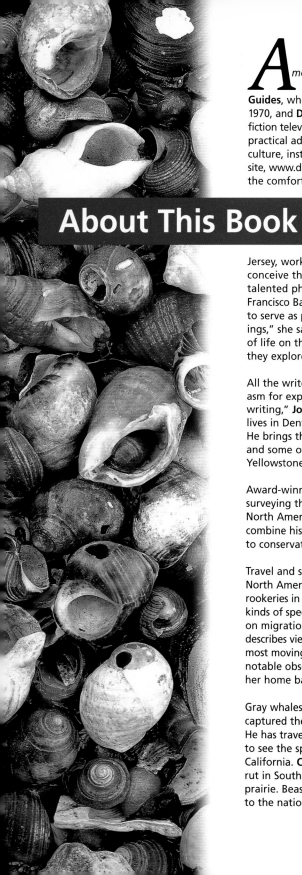

American Safari combines the interests and enthusiasm of two of the world's best-known information providers: **Insight Guides**, whose titles have set the standard for visual travel guides since 1970, and **Discovery Communications**, the world's premier source of non-fiction television programming. The editors of Insight Guides provide both practical advice and general understanding about a destination's history, culture, institutions, and people. Discovery Communications and its website, www.discovery.com, help millions of viewers explore their world from the comfort of their home and encourage them to explore it firsthand.

About This Book

This book reflects the contributions of dedicated editors and writers familiar with the top wildlife-watching destinations in North America. Series editor **John Gattuso**, of Stone Creek Publications in New Jersey, worked with Insight Guides and Discovery Communications to conceive the series and selected the stunning images by a roster of talented photographers. Gattuso called on **Judith Dunham**, a San Francisco Bay Area writer and editor with numerous books to her credit, to serve as project editor. "Observing animals in their natural surroundings," she says, "fills me with wonder and curiosity about the rich web of life on the planet. We want readers to feel that excitement when they explore this book while planning their travels."

All the writers share this fascination with wild animals and an enthusiasm for exploring North America's wildlands. "My whole aim in my writing," **John Murray** says, "is environmental awareness." Murray, who lives in Denver, Colorado, is the author of many books about wildlife. He brings this broad background to his chapters on observing wildlife and some of the many places he knows well: Coronado National Forest, Yellowstone National Park, and Alaska.

Award-winning writer **David Rains Wallace** was the ideal choice for surveying the diversity of animal life found throughout the habitats of North America. Wallace's many books, essays, and magazine features combine his astute observations of nature and his steadfast commitment to conservation.

Travel and science writer **Beth Livermore** has watched whales off the North American coast, observed manatees in Honduras, viewed penguin rookeries in Antarctica, and gone on safari in Tanzania. She knew well the kinds of spectacular wildlife-viewing opportunities to offer in her chapter on migration. One of these "big events" enthralled **Rose Houk**, who describes viewing sandhill cranes on Nebraska's Platte River as "among the most moving experiences I've ever had with wildlife." This is an especially notable observation coming from a writer who travels frequently from her home base in Flagstaff, Arizona, to write about nature.

Gray whales migrating along the west coast of North America have long captured the interest of writer **Peter Jensen**, who covers Baja California. He has traveled extensively in Baja for two decades, but is also fortunate to see the spouts of migrating whales from his home in Del Mar, California. **Conger Beasley** was drawn to another "big event," the bison rut in South Dakota, as well as to all the other creatures that share the prairie. Beasley has written more than a dozen books, including guides to the national parks of the Rocky Mountains and the Southwest.

Other writers describe places they have explored for much of their lives. **Janisse Ray**, who covers Everglades National Park and Cumberland Island National Seashore, is a native of the coastal plains of southern Georgia, to which she returned after living many years in Florida. Not surprisingly, she is most at home "in places where land and water meet." **Michael Furtman** feels a similar attachment to Boundary Waters Canoe Area Wilderness, which he began to navigate by canoe when he was a child. Living in Duluth, Minnesota, allows him to return often to canoe country. **Wayne Curtis** has undertaken expeditions for the Discovery Channel website covering tornadoes and icebergs. Other assignments, like writing about wildlife in Baxter State Park for this book and authoring guidebooks on New England and maritime Canada, have kept him close to home in Maine.

Bill Belleville started scuba diving in south Florida more than two decades ago and brings his vast experience to a chapter on Florida's John Pennekamp Coral Reef State Park. He's also filed dispatches while on expedition for the Discovery Channel website. Native Californian **Blake Edgar**, who covers the Monterey Bay National Marine Sanctuary, learned to scuba dive in the kelp beds of Monterey Bay and returns often to observe undersea life, study tide pools, and hike the shore. "Whether onshore or underwater," he says, "I always find plenty to intrigue me." **Glen Martin**, also based in the San Francisco Bay Area, writes about wildlife in one of his favorite places in California: Yosemite National Park. He has explored every corner of the park, seeking out the wildlife, kayaking the rivers, and backpacking the trails.

The southern Appalachian Mountain area is the turf of **Bruce Hopkins**, who lives near the Blue Ridge in northern Virginia and covers the wildlife of Great Smoky Mountains National Park, one of his favorite parks. He is also the author of guides to wildlife and Central Appalachia. **Mel White** has traveled close to his home in Little Rock, Arkansas, and far afield to write about natural history, including the rewards of birding, one of his favorite pursuits. For this book, he returns once again to Buffalo National River. **Pete Dunne**, author of astute essays on birding and birders, writes about Delaware Bay, a place he knows intimately and has observed season after season from his coastal perch at Cape May, New Jersey. **Lawrence Cheek**, based in Seattle and the author of several guides to the Southwest, tracks down the wildlife of Olympic National Park. "Living just a 30-minute ferry ride from the Olympic Peninsula," he says, "I constantly devise reasons to go there."

Thanks to Steve Manning of The Nature Company and the many park rangers and naturalists who reviewed the text. Thanks also to members of Stone Creek Publications' editorial team – Edward A. Jardim, Michael Castagna, and Nicole Buchenholz.

Tide pools (opposite) in East Coast sanctuaries like Acadia National Park, Maine, support a rich variety of marine life.

Gila monsters (above), venomous lizards, inhabit Southwestern deserts.

Barren-ground caribou (left), with their elegant, upswept antlers, are commonly seen by visitors to Alaska's Denali National Park.

Snowy owlets (following pages), born on the northern Arctic tundra, leave the nest after a few weeks but are still fed by their parents. Adult snowy owls feed on lemmings when in the Arctic.

Table of Contents

MAPS

ontemplate, for a moment, a dramatically different world, a place where wolves leave no footprints in the snow, where the music of songbirds never heralds the sunrise, where whales and dolphins never thread through the oceans. Along with envisioning barren continents and seas, you would have to recast the scenario of human evolution, for the lives of humans and animals have always been intertwined. Since humans first dwelled on Earth, animals were integral to their survival. As the imprint of *Homo sapiens* transformed the planet, the relationship between people and animals grew ever more complex and profound. ◆ Our enduring need for animals is as much spiritual as it is rooted in the practical necessity of feeding human populations. Their existence nourishes the soul and ensures a link with nature that is gratifyingly tangible, yet also **Experiencing wildlife** tantalizingly mysterious. An animal, wrote Helen **teaches us as much** Hoover about a fisher she watched in **about ourselves as about** the Minnesota woods, "adds joy to my days **the creatures with whom** because of its beauty and grace" and "stirs **we share the planet.** in me an eerie whisper of ancient hauntings." Beckoned from our habitat of urban grids and cement pathways, we journey to the wilderness in search of the creatures of the desert and mountain, forest and ocean. The process of watching an osprey seize in its talons a fish from a river, or a doe gently coax a fawn through an alpine meadow, sharpens our senses and heightens our awareness of the deep-seated kinship between humans and wild creatures. ◆ Animals, though they can be observed, never disclose every detail about their lives, and this is part of their allure. They instinctively follow rhythms and exhibit behaviors that even humans who live among

Ancient pictographs testify to a close bond between humans and wildlife.

Preceding Pages: Snow geese at Bosque del Apache, New Mexico; Rocky Mountains elk; walrus herd along the Alaskan shoreline.

tionships with all other life-forms, both animal and plant. Just as the bison of the Great Plains depend on the prairie dog, the prairie dog depends on the bison. As much as the corals of the Florida Keys need the parrotfish and cardinalfish, the fish need the corals. When surrounded by wildlands like the prairies and the coral reefs, we can rejoice in being vastly outnumbered by wild creatures, large and minute, bold and elusive, irascible and timid.

Unable to guarantee this biodiversity for themselves, animals rely on humans to guard it for them. North America is graced with wildlands, including those in this book, that sustain biodiversity by protecting an astonishing range of habitats. In the wetlands of the Everglades and on the tundra skirting Denali, from the ridge tops of the Olympic Peninsula and on the shores of Delaware Bay, you can be in the presence of wild animals in all their majesty, delve into their symbiotic relationships, and, in so doing, feel the quickening of your heartbeat.

them never entirely comprehend. They are, Edward Hoagland noted, "messengers from another condition of life, another mentality, and bring us tidings of places where we don't go." A hint of fear subtly infiltrates the desire to gain insight into the realm of animals. This sense of impendent danger was instilled long ago, when our ancestors preyed on wild animals while hoping not to be preyed on by them.

Listening to the yelping of

coyotes in the night or gazing, transfixed, on a rattlesnake sunning on a rock feels like a primal experience. "I have felt the inexplicable but sharply boosted intensity of a wild moment in the bush," Barry Lopez wrote, "where it is not until some minutes later that you discover the source of electricity – the warm remains of a grizzly bear kill, or the still moist tracks of a wolverine."

Revelations like this, at once inspiring and chilling, occur only when wild creatures are encountered in the wildlands where they live. *Wild* has many synonyms but no adequate substitute. A mountain lion in captivity still possesses beauty, even mystery. Yet the creature itself, along with our experience of it, seems less vital. Animals belong in places where they thrive in fine-tuned interdependent rela-

Encountering predators, even the small but fierce badger (above) underscores the instincts animals must possess to survive in the wild.

The northern saw-whet owl (left) is shy by nature but can be identified by its distinctive early-morning calls.

Nature's surprises can be subtle and strikingly beautiful, like this baby rattlesnake (right) warming itself on a calla lily.

SECTION ONE

◆

Preparing For the Field

◆

Advance planning helps enhance your experiences in the field. Now's the time to invest in binoculars, peruse field guides, learn wildlife-watching tips, and expand your knowledge of the places you will visit and the creatures you expect to see.

Observing Wildlife

When it comes to watching out for wildlife, keep in mind these three R's: the Right place, the Right season, and the Right time of day. They are the indispensable factors that can spell the difference between success and disappointment. So be prepared, especially if your trip involves travel to a distant or unfamiliar location. ◆ Be sure to plan ahead. As part of your preparation, get some friendly advice from the people in the park, refuge, and wildlife sanctuary offices. You can reach them by phone, letter, or electronic mail. Resident wildlife experts can direct you to the best viewing experiences, recommend the most productive times to visit, and help you better understand the significance of what you're apt to see. You'd be surprised at the amount of material they'll provide, free of charge – maps, descriptive illustrated brochures, fact-filled publications. Nowadays there are informative websites, often with visual images, maintained by the national parks and wildlife refuges. And you can

With the right equipment and the use of time-tested techniques, spotting wildlife is more productive and rewarding.

also surf the Internet for online resources related to your particular destination. ◆ Most professionals can recommend good field guides for birds, mammals, reptiles and amphibians, butterflies and insects that will help you prepare for your trip and identify species once you're in the field. Videotapes available at many public libraries can also provide insight into wildlife identification and behavior, and there are even tape recordings that preview songs, calls, and other sounds you can expect to hear. When you arrive at park or refuge, you may want to stop by at the headquarters or visitor center to check on that day's conditions. They're subject to change,

Brown bears feasting on spawning salmon draw wildlife watchers to Alaska's Brooks Falls in the summer months.

Red fox pups (preceding pages) born in spring are hunting on their own by fall.

especially where migratory species are involved.

Get a Good Look

Don't leave home without a good pair of binoculars, the simplest and most useful piece of equipment you can bring along. With them you'll be able to view from afar all manner of wildlife, from black-tailed prairie dogs to bighorn sheep, without disturbing natural patterns of activity or behavior. And for birding they're indispensable; few of us have eyes sharp enough to observe a lesser goldfinch or even a golden eagle at 100 yards.

Binoculars will also permit safe viewing of dangerous or shy fauna, like grizzly bears or many wild birds, for extended periods without scaring off the creature. For optimum viewing, you should have field glasses that enlarge an animal seven times what you would see with the unaided eye. Most seasoned wildlife observers prefer large binoculars over compact ones; the latter have smaller fields of vision and less light-gathering capabilities in low-light situations such as dawn and dusk.

For greater magnification (30 times or more), consider acquiring a spotting scope. Because of their increasing popularity, scopes have become more and more modestly priced. These high-powered miniature telescopes are especially valuable for viewing species found in inaccessible areas – isolated sea stacks, forest canopies, high cliffs – and for observing animals like the bull moose which, during the autumnal rut, may pose a threat to the viewer. Scopes are mounted on a stable tripod or, with a special attachment, on the window of a car. Daypacks permit you to carry scopes attached

Wildlife Ethics

Responsible wildlife viewing means knowing how to watch wildlife and when to give animals the space they need.

• Keep your distance and always carry a visual aid – binoculars, spotting scope, camera with telephoto lens. They permit safe viewing without intrusiveness.

• Be alert to what animals tell you. Their actions will generally indicate if you are too close or hanging around too long. Watch for salivating, pacing, jaw popping, sudden behavioral changes, or aggressive displays.

• Never approach predatory kill sites, natal den sites, bird nesting areas, or bear daybeds, digging sites, fishing areas, or hibernating dens.

• Do not take actions, from high-pitched whistles to herding, intended to move animals into a better position for viewing or photographing. Using calls can consistently attract animals but can also disturb them.

• Never feed or bait wildlife, and always pack out all food and garbage.

• Refrain from touching or lingering over a newborn animal in the field. A deer fawn's and elk calf's only defenses against predators are their lack of scent and their camouflage.

• Do not pursue animals in the winter on a snowmobile or even on cross-country skis. Every calorie is important at this time of year.

• Scrupulously adhere to all park, forest, and refuge regulations governing wildlife viewing.

• Know when you've seen or photographed enough.

• Respect the rights of others to watch the same animals. Take care not to block their view or disturb the peaceful silence that is so much a part of the experience.

Human trash (above) left in the wild can harm animals such as this coyote.

Posted regulations (left) aim to help protect both wildlife and watcher. Be sure to read them before entering a park or refuge.

Dall sheep, grizzly bears, and other Alaska wildlife can be safely observed and photographed from Denali National Park buses.

to tripods more conveniently and comfortably in the field than by simply slinging them over the shoulder.

Other simple tools can be of great practical service. A parabolic reflector, a directional microphone, and a headphone enable you to listen to even the faintest birdsongs, and can be used with recording equipment. Listening to such recordings will help sensitize you to the subtleties of songs and calls before you return to the field. For observing wildlife at night – usually the case when studying desert or beach fauna – you may want to cover the end of your flashlight with a red filter. Most animals do not perceive the red end of the spectrum as humans do and will proceed with their activities.

Keep a Low Profile

When you move about, remember to minimize your presence and avoid letting the animals know that you're there. Behave like the wild animals you observe. Wearing earth tones as much as possible will help camouflage your shape and color. Be quiet and move slowly when animals have their heads down so you don't startle them. Avoid standing conspicuously on a ridge or other high spot. In some areas, following game paths as well as human-made trails can bring you close to wildlife. By keeping the wind in your face, it's more difficult for an animal to smell you, and by keeping the sun at your back, you're more difficult to see. Take the time to stop for a while and just watch and listen.

Ecological hot spots are prime locations for spotting wildlife. These are places where water, shelter, and food regularly attract wildlife, such as a desert spring, a high mountain pass, or a berry patch. Every ecosystem also has its unusual hot spots. A walker moving through an old-growth forest in the Pacific Northwest at first might concentrate on natural clearings and streamsides. A dead Douglas fir, which may remain standing for as long as 50 years, rotting slowly from the top branches down, is a wildlife magnet for nesting creatures like spotted owls and raccoons. Time spent exploring these large snags and dead-

Debugging

Whether facing the stinging blackflies of Maine, the legendary mosquitoes of Alaska, or the infamous sand fleas of Georgia, you'll need to come prepared.

During the spring and summer months of peak insect activity, a head net is essential. These lightweight nets fit over most hats; some have clear mask-size windows that facilitate wildlife watching. For complete coverage, also wear heavy pants, a long-sleeved shirt, and cotton gloves. The strongest chemical repellents are those that contain deet as the chief ingredient. Also effective are repellents that use citronella, a natural ingredient.

Be aware of locations where you need to take measures to ensure your health as well as your comfort. Mosquitoes in some states, including Minnesota and Florida, can transmit the viral disease that causes encephalitis, a brain infection that can be fatal. Elsewhere, tiny wood ticks are known to cause Lyme disease. After being outdoors, check skin and clothes for insects and look for unusual bites that may need medical attention.

Head nets, plus an effective repellent, permit comfortable viewing of wildlife in habitats that also have species of biting insects.

did they come from and where are they going? Was the animal walking or running? Why was it here and what was it doing? What does scat, like that of a bear, or another sign, like an owl pellet, reveal about what the animal eats? How old is this bit of deer velvet (based on the drying of the blood) hanging from the maple branch? Pursuing these scattered clues can open a treasure-house of information about animals you may later see.

Insect watching may seem mundane when compared with the heroic efforts that must be undertaken to observe a mountain goat. But consider the example of Henry David Thoreau, entranced by watching a pitched battle involving two neighboring anthills. All the while he tried to determine the cause of the conflict, how the ants recognized one another, and what they did with the casualties. The most popular form of insect watching is observing butterflies and moths. Amateur entomologists can seek out a meadow full of nectar-producing blossoms to observe an array of butterflies on a summer day or see such spectacular moths as the large green luna and the giant cecropia on a summer evening. Most enthusiasts carry such essential tools as a local field guide (because

Analyzing tracks can reveal where an animal was going and what it was doing.

falls might prove more fruitful than time spent among the living trees. Similarly, a hiker above timberline may initially be drawn to search the wide-open expanses for wildlife. But timberline itself – the boundary between the treeless zone and the forest – actually provides the best opportunities for wildlife viewing. Such natural border areas, overlapping habitats called ecotones, support a rich concentration of food, shelter, and water.

Revel in the Details

Looking for animal tracks and signs will open a whole new world. Concealed in an otter slide beside a river, the daybed of a white-tailed deer, the marks of a bear's claws on an aspen tree are the stories of each animal's life. Sometimes as much can be learned by studying tracks and signs as by observing the animals themselves, especially those that are difficult to see in the wild, such as wildcats or elusive forest creatures like the marten and fisher.

As you analyze tracks or signs, ask yourself questions like these: What type of animal made the track? (Field guides devoted to tracks can help you identify them.) How old are the tracks? How many animals were there? Where

of the sheer number of insects in even a small area), a magnifying glass, and tweezers for examining specimens found dead in the field.

Take Notes

A final valuable tool for the wildlife watcher is a field journal, a diary containing details of your observations of animals and their environment. Naturalists from Aristotle to Lewis and Clark to Edward Abbey have maintained carefully written records of their observations. Some have also incorporated sketches and drawings. Others have included photographs with their writings.

Whether you record your observations for future publication or for personal reference, journals help you remember important details

about your experiences in the outdoors. The very act of writing what you see, hear, and smell deepens your understanding of nature. Without a written record, even key facts such as the appearance and behavior of the species and the time of day it was seen can easily

be forgotten. The value of a field journal increases over time, as you'll see when you later read it to reflect on your experiences or prepare for another journey into nature.

A telephoto lens (above), as well as binoculars and spotting scopes, brings an animal close enough for a frame-filling photograph.

Inquisitive animals like this sandhill crane (below) occasionally approach a photographer for an up-close view.

Photo Op

Almost everyone who observes wildlife carries a camera. With each year, 35 mm cameras become more powerful, portable, and inexpensive. Depending on your destination, you may also want to bring an underwater camera or even a video camera or camcorder. Wildlife still photography is not difficult, so long as you apply some general principles.

• Using a tripod is advisable, even if you are just stepping from the car to photograph a band of pronghorn antelope from the road. A tripod stabilizes the camera, thereby producing a sharper image. More important, it forces you to concentrate and create better compositions. A tripod is a necessity when using a telephoto or zoom lens.

• Determine the type of film you want to use. Does slide film or print film best meet your needs? Most wildlife photographers prefer to shoot color, though black and white may be appropriate in some situations.

• Film speed is an essential consideration. Do you want 400 ASA film, which is easier to use in low light but produces grainy images? Or, like most professionals, do you want more demanding 64 ASA film, which is difficult to use in low light but produces sharper images? You may want to bring film in a variety of speeds for different light conditions.

• Make sure you bring more than enough film. It's easy to carry extra rolls, but when your supply runs out, it's hard to find film in the backcountry of a wildlife refuge.

• As all great wildlife photographers discover, patience is a virtue. Even in an area where animals are plentiful, you may need to wait for hours, even days, for the weather to clear, or for the light to be just right, or for animals to begin interacting in interesting ways.

Habitat Guide

From arctic tundra in Alaska to subtropical hardwood forest in the Florida Everglades, North America encompasses many habitats – the places where animals and plants live. Black bears, for instance, usually seek out woodland; grizzly bears tend to prefer more open areas such as grassland. The causes of such preferences are complex. One probable reason black bears live in woodland is that they mainly eat plant foods like nuts and berries. Grizzlies also eat mostly plants, but sometimes prey on large animals such as elk, which form large herds in grassland.   ◆   Most North American habitats are classified as "temperate," though they may undergo tropical heat in summer and arctic cold in winter. Seasonal changes in weather and vegetation channel activity into predictable patterns or cause shifts in habitat preference. A black bear may emerge from the **The more you know** forest to feed on prairie grass in spring; a grizzly **about animals and their** may move into the woods in fall when **habitats, the easier they are** nuts and berries are ripe. Habitats tend to be **to find and understand.** mingled, as when streamside woodland and wetland occur with grassland. Wildlife take advantage of such habitat "mosaics" to get the best of several worlds, as when both black bears and grizzlies frequent streams during fish spawning runs.   ◆   Wildlife activity is bewilderingly diverse, but some rules apply. Daily activity usually peaks at dawn and dusk, when most animals are seeking food. Yearly activity peaks at times of seasonal change, when animals are breeding or migrating. Animals often move between habitats, making ecotones, or edge environments, good observation sites. Reviewing the following pages will help you understand how animals live in a variety of North American habitats – which is essential for observing them.

Mountain goats seek out precipitous slopes at treeline from the Northwest through western Canada and into Alaska.

SEASHORE

Wildlife is intense at the original edge environment: the seashore. Marine creatures such as whales and seals often surface just offshore, while abundant food along the shore attracts land animals. On parts of the West Coast, sea otters feed on mollusks in floating kelp beds, and river otters fish in adjacent estuaries. Many fascinating creatures are neither strictly land nor sea animals but are specialized for a life between the two.

The sea causes characteristic patterns of wildlife activity. Tides are as influential on daily activity as dawn and dusk. Mussels, barnacles, and oysters open their shells to filter food from the rising water, then close them as the tide recedes. Shorebirds congregate on tidally exposed reefs and sandbars to feed, then fish replace them with the incoming tide, sometimes hunting the same small organisms as the birds. Changing tides give human visitors access to the seashore world – as tide poolers when the water is

out, as sea kayakers and snorkelers when it returns.

In autumn and winter, the sea's moderating climatic influence attracts vast migratory bird populations. Dozens of shorebird species frequent beaches, from tiny least sandpipers to willets and yellowlegs. Gulls, terns, and pelicans use beaches as rest stops between fishing forays; other fish-eating migrants such as loons, diving ducks, and murres swim offshore. On bays, herons and other wading birds dot the waterside, and secretive rails lurk in marshes. Bald eagles, ospreys, harriers, and short-eared owls also congregate on coasts in winter. Some North American coastal migrants are spectacular and rare, like the whooping cranes that winter on the

barrier islands of the Texas Gulf Coast.

Birds aren't the only migrants visible on winter shores. Gray and humpback whales have been an attraction on the West Coast for decades as they move between tropical breeding areas and arctic feeding waters. Migrating whales are increasingly numerous on the East Coast as populations recover from the ravages of the whaling era. Another

Great blue herons (left) inhabit wetlands, lakeshores, and tidal flats well supplied with fish, frogs, snakes, and aquatic insects.

Breeding California sea lions and northern fur seals (right) gather on the Channel Islands off the Southern California coast.

Blue whales (left) reach 100 feet when fully grown, making them the largest known animals. Newborn calves are 23 feet long.

Starfish (right) are often found clinging to rocks, especially in tide pools at low tide.

spectacular migrant is the northern elephant seal, which breeds on the central California coast in winter. Elephant seals spend most of their lives feeding in deep Pacific waters and put on so much blubber that they don't eat at all during their several months of breeding on the beach.

Coastal bird activity decreases in spring and summer as migrants depart to the interior or Arctic, but there are many exceptions. East Coast barrier islands and West Coast offshore rocks support seabird nesting colonies, as do some isolated cliffs and beaches. Observing

such colonies requires effort and care, which is a good thing since the birds wouldn't be there if they were easily accessible. Organized tours, such as the whale-watching trips that circle the Farallon Islands off San Francisco, are often the best ways to see breeding colonies and may include unusual sightings like the huge leatherback turtles that visit the North American coast to feed on jellyfish.

Seashores teem with animal life even when deserted by birds and mammals, although it may be inconspicuous or secretive. Burrowing clams, crabs, and worms honeycomb apparently empty beaches, sandbars, and mudflats, and their activities can be observed with a little patience. Breaking surf on southeastern beaches uncovers thousands of tiny, rainbow-colored coquina clams, which protrude their translu-

cent "feet" to dig back into the sand. On rocky shores, a few minutes of watching a tide pool can be even more rewarding as what first seems a mere heap of stones and seaweed gradually resolves into a menagerie of hermit crabs, sea stars, anemones, urchins, and, perhaps, a shy, camouflaged octopus crouched in a mussel bed.

Such "lower" animals may seem static compared with birds, but many make epic migrations at some stage in their lives. The most spectacular is that of horseshoe crabs, which move to mid-Atlantic shores to breed in spring. Walking a beach at dusk may reveal masses of their prehistoric-looking carapaces at the water's edge. Their eggs are a major food source for northward-migrating shorebirds.

WETLANDS

Despite a reputation for gloominess and inaccessibility, wetlands tend to be spacious and full of light and color. It's true that you have to be careful not to get lost in their mazes of waterways, yet travel can be enjoyable and relatively easy on the canoe trails that thread

River frogs, true to their name, live along slow-moving creeks near wetlands. Graced with long legs, they are powerful jumpers.

through vast areas like the Okefenokee, Everglades, and Boundary Waters. Wildlife activity often is great because, like seashores, wetlands exist on the fertile boundary of water and land, and some of the rarest and most beautiful organisms inhabit them.

Wetlands occur wherever land is flooded for at least part of the year, from seashore salt marshes to river swamps and lake marshes to mountaintop bogs and fens. Even deserts have wetlands around oases, river sinks, and salt lakes. Wildlife varies according to location, but

Wood ducks nest in tree holes near water. The male is considered one of the world's most beautiful ducks.

some specialized wetland organisms are characteristic. Cattails, sedges, and rushes predominate in marshes, while water-tolerant trees like bald cypress and box elder dominate swamps. Growing in bogs and fens are strange pitcher plants and sundews, carnivorous plants that trap insects to supplement their nitrogen supply. Reptiles and amphibians are particularly common, moving easily between land and water. The alligator rules wetlands from Virginia to Texas. This ancient creature that survives elsewhere only in southern China helps to maintain its habitat by digging "gator holes" that harbor other wetland species during droughts.

Wetlands are most exciting in spring. Many water birds that winter on seashores nest in swamps and marshes, good places to observe their breeding plumage and behavior. Sandhill cranes dance and circle high in the air above north-

Saltwater marshes, lakeshore swamps, and alpine bogs support animals of many species, from reptiles to birds, that need to live in or near water.

ern and western marshes. Snipe and woodcock perform intricate twilight rituals in eastern swamps and bogs. Wetland specialists like wood ducks and prothonotary warblers are among the most colorful North American birds in spring plumage.

Spring reptile and amphibian activity may surpass even that of birds. Bull alligators roar to attract females, which lay their eggs in nests of rotting vegetation and carefully guard them. Frog and toad calls can be nearly deafening, with many different species calling in eastern wetlands. Even subarctic muskegs resound with frog calls in spring because the native species, the wood frog, has somehow evolved the ability to survive freezing solid in winter. Masses of gelatinous eggs in pools attest to amphibian fertility. Salamanders such as newts and ambystomas deposit eggs, as do frogs, and you'll be able to see them if you walk the swamp margins with a flashlight at dusk. Frogs may stop calling and duck underwater as you approach, but they will start again if you stay still a few minutes. Their vision is sensitive mainly to movement.

Spring is also a good time in wetlands because biting insects haven't reached their peak. Wildlife becomes less active in summer anyway, as birds secretively fledge young and then undergo their annual molt. Pools shrink, and amphibian larvae grow legs and disperse to land. Wading birds such as bitterns and wood storks often congregate around shrinking pools to catch tadpoles and fish. Whitish fragments on banks show where turtles and snake eggs have hatched or been eaten by raccoons.

Activity picks up again in late summer and fall. Cardinal flower, aster, gentian, and other colorful wetland wildflowers bloom. Migrant birds gather into large flocks, like the clouds of southbound swallows that billow over marshes. As the days shorten and the autumn wind bends rushes and cattails, the calls of geese and cranes evoke a sense of wanderlust. The year's crop of young muskrats, minks, and other marsh mammals sets out in search of a winter home.

Wetlands can be lively even in winter. Where there is open water, waterfowl may be as abundant as on the shore. The marshes of California's Central Valley and the Mississippi Delta teem with snow geese, some arriving from as far away as Siberian nesting grounds.

High water levels can make for especially good canoeing at this time. Even frozen northern swamps and bogs can provide good snowshoeing and cross-country skiing.

Beavers (above), once hunted to near extinction, thrive in wetlands throughout North America. Listen for the slap of their broad tails as they dive underwater.

Parks and refuges in south Florida preserve habitat essential for the endangered American crocodile (below).

WOODLANDS

Forests are both familiar and strange. A walk in the woods is a basic way to observe wildlife, and reveals much, but parts of the forest such as the canopy and underground remain little known. With its maze of foliage, a great forest can seem wilder than almost any other habitat.

Largely wooded when humans arrived, North America retains the world's richest temperate forests, with the greatest diversity of tree species, from western redwoods to eastern Appalachian hardwoods. Four levels of plants characterize old-growth forests: a canopy of big trees, a subcanopy of smaller trees, an underbrush of shrubs, and a ground layer of herbs. Wildlife is correspondingly diverse, although the overwhelming vegetation can make it harder to see than in habitats that are more open. Forests may seem devoid of animals at times, but this is usually an illusion, and their presence is often

detectable by signs, sounds, and even smells. At other times, animals may be visible everywhere.

In eastern deciduous forest, the first warm days of spring bring a burst of activity as animals search for mates and food. Hibernating frogs, box turtles, chipmunks, squirrels, groundhogs, and black bears emerge, and migrant songbirds arrive. Trillium, hepatica, trout lilies, and dozens of other wildflower species carpet the forest floor. Since tree leaves aren't yet unfurled,

you can get an unobstructed view of activities that foliage will hide for the rest of spring and summer. Early spring also is lively in western and northern coniferous forests, although evergreen foliage makes the activity less conspicuous.

Throughout spring and early summer, the chorus of songbirds at dawn and dusk is a source of fascination and sometimes frustration. Foliage often hides the singers, but you can attract singing birds by imitating the scolding calls of chickadees

Porcupines often climb trees, sometimes all the way to the top. They like to munch on the inner bark and tender buds.

and titmice. Curious songbirds sometimes alight a few feet away, allowing you to match songs with singers, from the pewees that whistle or buzz in dawn's first light to the thrushes that sing as the evening stars emerge.

Many forest creatures sing because it is an important way to locate mates and claim territory among the trees. Whip-poor-wills and chuck-will's-widows call through summer nights in eastern forests, and owls can be heard all year, from the whistling of screech owls to the deep booming of the great horned owl. Insects' calls are as diverse as those of birds, and become particularly strident in the heat of high summer. Cicadas and grasshoppers whir through afternoons, and katydids and crickets begin at dusk. Insect calls are ventriloquial, so callers are hard to locate, but patient examination of foliage often reveals a bizarre creature rubbing its wings together to make a noise like a live wire or a tiny bell.

Mammal activity becomes conspicuous in fall, when

Deer mice (opposite), common throughout North America, gather seeds, acorns, and other nuts and store them in their woodland caches.

Caterpillars (right) tend to have robust appetites. Some feed on specific host plants, while others have extremely broad diets.

many species mate. Buck deer spar in woodlands everywhere, and bull moose somehow navigate their enormous antlers through northern forests. Signs of the deer rutting season are easy to find in saplings splintered by antlers or ground trampled during sparring matches. Animals hurry to put on fat for the winter, and groves of oaks, beeches, hickories, and other nut-bearing trees attract congregations of squirrels, raccoons, and wild pigeons, as well as bears and wild boars. Watching quietly for a morning or evening is a good way to experience the interface between the treetops and forest floor.

Early fall is an active time for the most abundant vertebrates in temperate forests: lungless salamanders. They mostly live underground, but on warm, rainy nights in September and October, a walk in the woods will often reveal hundreds of these small amphibians wandering over the leaf litter.

Unlike newts and amby-stomas, many lungless salamanders lay their eggs on land. Dozens of species inhabit North America, from wormlike slender salamanders of western forests to colorful red-and-green salamanders of the East.

Forests become quieter in winter after migrants leave and hibernators retire, but many resident birds and mammals stay active. Congregating for safety from hawks, mixed flocks of woodpeckers, nuthatches, titmice, creepers, and chickadees move among the bare trees picking insect cocoons from the bark. Opossums creep over the ground or climb vine tangles, sometimes with frostbitten ears and tails in northern woods. Snow provides an easy way of tracking the behavior of secretive creatures – the brilliant white carpet after a storm can record a fox's search for food or an owl's snatching of an unwary mouse from the ground.

GRASSLANDS

Grassy plains may seem barren compared with lush woodlands. Early European settlers bypassed the Midwest's tallgrass prairies thinking they would not support farms. In fact, the opposite proved true. Grasslands are among the most productive habitats, particularly for large mammals such as the bison, pronghorns, grizzlies, and wolves that once thronged the plains.

Farming has preempted most North American grasslands today, and specialized creatures such as the black-footed ferret and swift fox are rare. Wild bison still live in places like South Dakota's Badlands National Park, however, and much wildlife persists in farmed areas. Although pronghorns were nearly exterminated in the 19th century, they can coexist with ranching and have reoccupied much of their original range from Manitoba to New Mexico. Small herds are commonly seen along interstate highways, but getting a closer look is another matter, since they run extremely fast. They are curious, however, and sometimes will approach to examine strange objects.

Because of their exposed, continental location, grasslands have extreme seasons. Unchecked winds intensify the effects of heat and cold. On the arid high plains, wind and water erosion has carved large areas into the strange buttes and spires of badlands, where few plants can grow on the alkaline soil.

Bison, an emblem of the Great Plains, once roamed native grasslands in large herds.

In pre-farming days, grasshopper swarms devoured the prairie for miles, and fires burned for weeks during the dry times of fall and early spring. Bison and other animals sometimes died by the thousands in droughts, floods, and long, hard winters. Snow still drifts high in blizzards, and warm-season thunderstorms and

Greater prairie-chickens engage in a fascinating mating ritual. The male attracts a mate by emitting a low booming call with colorful air sacs called tympani.

tornadoes can be menacing.

Conditions are more stable underground, and prairie grasses and forbs have most of their biomass in the soil as roots or bulbs, safe from drought, frost, and wind. Such plants can live many years, with only a few leaves showing above the surface. Many grassland animals are burrowers, including the prairie dogs whose "towns" once extended miles across the high plains. Prairie dogs are related to squirrels, and many other ground-dwelling squirrel species inhabit North American grasslands. Their colonies provide habitat for other creatures. Badgers, snakes, and black-footed ferrets prey on the rodents, and burrowing owls, box turtles, toads, lizards, and salamanders find shelter in the burrows. Observing a ground-squirrel colony is a good way to experience the concealed diversity of grassland wildlife.

Despite climatic extremes, grasslands make for exhilarating hiking. Spring wildflowers are spectacular, like the "bluebonnet" lupines that paint the Texas prairies.

Pronghorns (left), the fastest mammals in North America, can reach a top speed of 60 miles an hour.

Burrows dug by prairie dogs (right) are also used by burrowing owls and black-footed ferrets.

Nesting birds include bobolinks, dickcissels, and upland sandpipers that migrate from South America, as well as resident meadowlarks, horned larks, and prairie chickens. The "pothole" wetlands of the glaciated northern plains attract surprising numbers of nesting waterfowl, wading birds, even seabirds like gulls, terns, and white pelicans.

Summer grasslands tend to be bleached by the sun and parched in the region of the upper plains and Far West, with only aromatic tarweeds and vinegarweeds blooming. In the humid Midwest, late summer is the peak of flowering, when tallgrasses like big bluestem grow over your head, and showy forbs like royal catchfly, compass plant, and queen of the prairie reach even higher. Butterflies and bees swarm among the giant wildflowers, and equally colorful crab and jumping spiders lie in wait for them, while thousands of orb webs glitter in morning dew. Large tallgrass prairies remain in only a few places, like Tallgrass Prairie National Preserve in Kansas, but patches survive in regional parks, pioneer cemeteries, and railroad rights-of-way as far east as Ohio.

Wild grasses are highly nutritious, and some species remain so even after they dry out in fall, "curing on the stem," as ranchers say. This rich food supply allows some animals to stay active even through harsh winters in the Great Plains. Bison plow through snowdrifts to reach fodder, impervious to cold under their shaggy coats. Pronghorns find grass on windswept ridges. Voles and gophers scurry through burrows and runways under the snow, and coyotes occasionally can be seen rearing up on their hind legs to pounce on hidden rodents.

DESERTS

Deserts are among the best habitats for watching wildlife, which may seem paradoxical considering their harsh climate. A surprising diversity of organisms has adapted to aridity, escaping the stiff competition of lusher environments.

Deserts often seem lifeless in mid-summer noon heat, but wildlife is active and easy to observe at gentler times.

Small animals like rodents and reptiles do particularly well in deserts by minimizing food and water needs. Lizards and pack rats can get all their water from food juices, and kangaroo rats metabolize it from a diet of seeds. Many desert creatures thus thrive far from water holes and water-drinking predators, and tend to be less wary than their grassland or woodland counterparts. If you watch quietly at a kangaroo rat colony or pack rat nest, you may see the occupants on morning or evening errands. You almost certainly will see lizards: vegetarian chuckwallas sunning themselves on stones, horned lizards lying in wait for ants, or collared lizards running about on their hind legs like little dinosaurs.

Deserts are by no means devoid of large wildlife. Pronghorns, bighorn sheep, deer, and javelinas thrive in arid country where streams or springs are accessible. Mountain lions, bobcats, and kit foxes are typical predators, and tropical creatures like coatimundis, ocelots, and jaguarundis are found near the Mexican border. Golden eagles and prairie falcons nest on cliffs, and handsome black, white, and chestnut Harris' hawks may nest atop saguaros.

Desert plants also thrive by minimizing water needs in ingenious ways. Cactuses and paloverdes dispense with leaves and photosynthesize through green stems. Tiny creosote bush leaves are covered with drought-resistant resin. Desert vegetation can be lush in years of high rainfall, however. Shrubs that are usually bare, like ocotillo, leaf and flower almost overnight, and wildflowers carpet ground that seemed forever sterile a few weeks before. Seeds can lie dormant for many dry years, then bloom all at once in a wet one. Animals respond promptly to spring rains. Nectar-eating bats migrate north from Mexico to feed on the large cactus flowers, thus helping to disperse their pollen. A variety of amphibians – spadefoot toads, desert treefrogs, tiger salamanders – emerge from underground to breed in temporary pools. Pupfish spawn; these are colorful minnows adapted to waters too hot or salty for other fish. Some pupfish species survive the complete drying up of their tiny pools or streamlets as eggs in mats of vegetation.

Nesting birds abound in

Greater roadrunners (above) rarely fly. Instead, they run across the desert to catch lizards, snakes, insects, and even other birds.

Desert tortoises (left) dig burrows to escape the desert heat, ambling out occasionally to feed on plants.

Cardon cactus, elephant trees, and flowering brittlebush (left) provide shelter, nesting sites, and food for creatures in Baja California.

Reptiles like this yellow-headed collared lizard (below) have adapted to life in the desert by obtaining water from their food sources.

desert, and spring mornings are loud with the cackling of cactus wrens and the four-note call of Gambel's quail. Many birds inhabit holes in large plants like Joshua trees and saguaro cactuses. Watching such holes may reveal woodpeckers or elf owls ferrying food to nestlings. The ground-dwelling roadrunner prefers to nest in the spiniest of cactuses, the spiny cholla. If you happen to park near a roadrunner nest, you may well see one of the wily birds slink out of the brush to pick smashed insects from your vehicle's radiator.

The desert spring is short. Activity subsides in midsummer, then picks up again later in the season. In the Sonoran Desert near the Mexican border, tropical storms from the Gulf of Mexico bring downpours that turn dry washes into torrents. Wildflower carpets and amphibian choruses can occur in August and September. Strange desert creatures like tarantulas, solpugids, and whip scorpions become noticeable in late summer as they move about in search of mates. Tarantulas look frightening, but most North American species are less dangerous than bees. Big bluish, red-winged wasps, often seen walking over the ground at this time, have a worse sting, but they use it to paralyze tarantulas, upon which they lay their eggs. After they hatch, the larvae feed on the big spiders.

Autumn and winter can be chilly, especially in high deserts like the Mojave and Great Basin. Temperatures drop swiftly in the dry air, and snow is not infrequent, although it seldom accumulates. Some animals such as ground squirrels and desert tortoises hibernate, and there is even a bird – the poorwill, a western relative of the night-calling whip-poor-will – that spends cold weather torpid in rock crevices. Most warm-blooded animals remain active, and even lizards may come out on sunny winter days.

MOUNTAINS

Most of North America's wilderness areas are in mountains, so the wildlife there tends to be spectacular. Some species are specialized for the rocks and ice of high-elevation country. Many more thrive on slopes and foothills. Mountain solitude offers hikers a chance to get close to wild animals.

Altitude and geology pose special problems for alpine life. Temperatures can be subarctic at high altitudes, air is thin, and exposure to sun and wind makes climate volatile. Gravity is a powerful force on steep slopes, causing landslides and molding snowfields into glaciers that flow gradually downhill, crushing everything in their path. In the West and Hawaii, active volcanoes can erase every organism from their vicinity, as with the 1980 eruption of Mount St.

Helens. These earth-shaping forces make for sublime scenery, and wildlife recovers from such cataclysms with surprising speed.

Despite a violent volcanic past, western midelevation regions like Yellowstone support North America's most diverse population of large mammals. The last free-roaming bison south of Canada found refuge in Yellowstone's valleys in the 1800s. The Rockies remain the grizzly's final refuge in the lower 48 states. All the West's native large grazing animals survive, and reintroduction of the gray wolf to Yellowstone in 1995 restored a full diversity of presettlement predators. Much large wildlife also lives in eastern mountains, such as black bears and cougars, or mountain lions, found in the Great Smokies.

Large mammals are less prevalent in high-elevation rock fields and tundra, but grizzlies live at such heights, and bighorn sheep and mountain goats can negotiate craggy places. Smaller mammals like marmots and pikas abound, often around mountain lakes. Pikas, short-eared rabbit relatives, live on rocky slopes, piling up grasses and wildflowers in miniature haystacks. They remain active under the snow all winter, whereas marmots, which are related to woodchucks, hibernate. Both marmots and pikas make shrill calls when intruders appear, a warning against the foxes, martens, fishers, and wolverines that hunt in the high country.

Mountain spring comes late, especially in the West where snow accumulates to 10 feet or more. Frogs and toads sing around high lakes and streams even before snow melts. Many common migratory songbirds nest in mountains, however, and some prefer them, such as bright-yellow-headed hermit warblers and tiny calliope hummingbirds. Altitudinal migration is as important for mountain creatures as latitudinal, and many species winter in valleys and foothills, then move upslope for breeding. Seasonal vegetation

Mountain lions (left) are stealthy and efficient hunters that prey on deer and smaller animals such as rabbits and rodents.

Bighorn sheep (right) graze alpine meadows or rocky slopes. The male's massive coiled horns are easy to spot.

regimes shift dramatically at different elevations, from the early-leafing hardwoods of foothills and lower slopes to the conifer forests of midelevations to the brush, meadows, and tundra of high elevations.

Pikas, found at high altitudes near timberline in western mountain ranges, build haystacks of food to survive the winter.

Short summers make for intense summer wildlife activity as animals hurry to raise broods and put on fat before fall, a rewarding situation for watchers. Butterflies and moths are abundant and diverse in flowery mountain meadows, sometimes as migrants, sometimes as residents. Morning frost is common throughout the summer above 7,000 feet, but high meadow wildflowers such as paintbrush, bluebells, sneeze-weeds, and gentians have adapted to it. Fall comes early, with western aspens and cottonwoods turning yellow in September. The "bugling" of bull elk echoes throughout the Rockies and Northwest. Autumn hardwood foliage in eastern mountains is the world's most colorful.

Winter descends with awesome speed, and snow-storms can hit as early as September. Many creatures remain active in the high country in winter, however, like the Clark's nutcracker, a western crow relative that caches seed supplies which allow it to survive extreme weather. Cross-country skiing and snowshoeing provide the greatest opportunities for observing wildlife at this time of year.

CORAL REEFS

Coral reefs are among the oldest habitats. Tiny jellyfish relatives called polyps make reefs by secreting protective limestone around their bodies, thus forming the planet's largest animal-built structures. Reef corals depend on photosynthetic algae living in their tissues for nourishment, and the algae require warm, clear water, so reefs occur largely in the tropics, although some are in southern Florida.

Since reefs are tropical habitats, their wild-life activity never seems to subside. An extraordinary diversity of organisms has evolved, making them unforgettably strange and colorful.

Reefs are among the most accessible marine habitats, because they mostly occur in shallow water, A pair of swim goggles is all that's really needed to explore them, although snorkeling or scuba gear allows an even more intimate acquaintance. Glass-bottomed boats like those serving John Pennekamp Coral Reef State Park in Key Largo also provide access.

Unfortunately, damaged reefs can turn quickly from wonderlands to wastelands. Although seemingly inexhaustible in their color and diversity, they are highly vulnerable to environmental changes. Hurricanes and human disturbance can kill corals very quickly. Since reef waters are relatively poor in nutrients, coral destruction spells disaster for fish. Existing at the tropical margin, Florida's reefs are particularly vulnerable to this sort of damage.

Reef corals come in bizarre shapes and sizes, from treelike elkhorns to boulderlike brain and starlet corals to little finger and flower corals. Many non-reef-building corals occur, too, in featherlike or fanlike shapes.

Reef denizens have evolved innumerable ways of life in this complex environment. Iridescent green or

Many species of fish depend on coral reefs, fragile structures formed over hundreds of years by small creatures called polyps.

A convictfish (left) hides in the tentacles of an anemone to escape predators. The fish is not harmed by the stinging tentacles.

The large red eyes of a small fish called a spotjaw blenny (below, left) scare away its enemies.

Sea turtles (below, right) swim along shallow reefs, feeding on algae.

blue parrotfish bite off chunks from the living reef, digest polyps and algae, and excrete the rest as white sand. Filefish, trunkfish, angelfish, and triggerfish of every imaginable color nibble on corals and sponges or pick small invertebrates from the reef. Loggerhead and hawksbill turtles also eat sponges and jellyfish. Big schools of blue tangs, grunts, and snappers throng open water. Spiny lobsters and red, goggle-eyed squirrelfish are among many creatures that hide in crevices by day, feeding at night.

The large predators attracted to this abundance are magnificent. Big reef fish may become tame and wait under dive boats for handouts, but, like bears, they are better left unfed. Groupers, nurse sharks, and moray eels lurk on the bottom. Bull and lemon sharks, barracudas, snook, and tarpon cruise open water. Prey have developed some ingenious defenses. Many fish have camouflage markings or eyelike spots on their back ends to confuse attackers. Puffer and porcupine fish inflate into prickly spheres when molested, and may have poisonous flesh, a common trait of reef fish.

Birds and mammals are less in evidence than fish on reefs, but are common in adjacent habitats. Manatees graze in sea grass flats, and dolphins hunt over sandy plains. In Hawaii, the endangered monk seal occurs around reefs, as do humpback whales. Cormorants, pelicans, and some tropical seabirds feed on reefs and nest on the islets called keys. Shorebirds and songbirds pause on keys during migration, picking up crustaceans or worms to fuel their journeys. Interesting creatures like land crabs and lizards live on them full-time.

A Gathering of Creatures

octor Doolittle spanned the globe to spy on wild
animals, but you don't have to leave the continent to do the same thing.
North America, between the icy Arctic and sultry tropics, has some of the
world's most spectacular migrations. Bats, sea turtles, polar bears, and other
creatures fly, swim, and wander ancient pathways across this mighty continent, and you can catch them in the act.   ◆   Strictly speaking, migration
is defined as a regular, seasonal journey made by animals to another habitat.
More broadly, it is the movement of any organism from one place to another
to find food, rear young, or meet other basic needs. The journey can be
triggered by internal cues, such as hunger or surging reproductive hormones.
Environmental cues, like shifts in weather, length of day, or phases of the
moon, may also tell animals that it's **Obeying a seasonal time clock,**
time to move on. Most often, the impetus is **herds of caribou cross**
a combination of the internal and the envi- **the tundra, flocks of cranes**
ronmental.   ◆   North America is perfectly **take to the sky, and**
situated to host a fantastic variety of migra- **grizzlies feast on salmon.**
tions. It has an assortment of climates and habitats that can accommodate
long-distance migrants as they travel north and south. As the Earth tilts
toward the sun in spring, or away from it in fall, these ecosystems are
transformed. Native species trade mountains for tundra, dens for riverbeds,
and ocean bottoms for beaches.   ◆   Indeed, migration season is generally
the best time to spy on animals for several reasons. Animals often travel in
groups, so it is easy to see huge numbers together in one place. Caribou tend
to winter in small scattered groups throughout Canada and Alaska, but in
summer they amass in great thundering herds that course across sprawling

Sockeye salmon crowd a creek in Alaska
as they swim upstream to spawn. Along
the way, they will be preyed upon by
black and brown bears and bald eagles.

Musk oxen, native to northern Canada, inhabit low-lying valleys in summer, then migrate to windswept mountains where they graze on plants not buried in snow.

grassy plains. Elsewhere, 80 percent of the world's sandhill cranes stop over at Nebraska's Platte River in spring. At twilight, thousands of the twiggy birds stand shoulder to shoulder in the placid river shallows where they rest.

Since animals move to new territory during migration, this is a good opportunity to spot otherwise aloof and exotic creatures. Polar bears, which inhabit the pack ice of northern Canada most of the year, are forced south in summer and remain on the shores of Hudson Bay near Manitoba until freeze-up. There, some 20 adults, weighing up to 1,200 pounds each, are visible from the shoreline. In summer, Mexican free-tailed bats cross the border by the millions to hunt moths throughout the southern United States. Even though you may never travel to Mexico, you can see one of its fascinating native creatures.

Because animals migrate to meet basic needs, like breeding and feeding, there is always plenty of action. On Pacific Coast islands, male elephant seals gather each winter to do bloody battle. The winning bull earns the right to mate with every female in his newly established territory. Loggerhead turtles drag their tired bodies onto Florida's barrier beaches each summer to dig holes with their back flippers and bury clutches of eggs. Two months later, the hatchlings emerge en masse and scurry for the sea.

Sometimes, two migrations coincide, bringing together prey and predator. The result is a dramatic feeding frenzy. Each spring, thousands of horseshoe crabs emerge from Delaware Bay to spawn on the adjoining shoreline. At the same time, 1.5 million migrant shorebirds, the second largest concentration in the world, stop by to rest on their way north. The birds slurp down about 9,000 eggs a day. In fact, each bird regains 50 percent of its body weight in two weeks or less. In Alaska, grizzly bears leave their dens in search of food at the same

Walrus (left) gather on Round Island in Alaska's Bristol Bay beginning in late spring, when the northern ice pack starts to melt.

Caribou (below, left) are nomadic animals that winter in small groups, then move in large herds to the arctic plains.

Buteos like the red-tailed hawk (below, right) are among the raptors seen heading south in fall.

time that salmon, which migrate thousands of miles in the open Pacific Ocean, arrive in freshwater rivers. This protein banquet is enough to fuel these burly bruins through the entire winter to come.

But that is not all. There are plenty of more subtle, yet equally inspiring thrills for the studious wildlife watcher. Observers of monarch butterflies are bound to discover that these feather-light insects can fly well over 100 miles per day, at altitudes of several thousand feet or more, fueled by flower nectar alone. By the same token, birders will learn that most birds have what amounts to a natural compass in their brains. They use this, along with the sun and stars, to navigate their longest, most adventurous flights.

Whether you want to know the best time to watch humpback whales breach like missiles, bison assemble into herds, or raptors float in the firmament, the following chart will tell you where and when to go. But be forewarned. Though wildlife events are as regular as the changing seasons, actual time frames can vary with weather and other less reliable factors. So be wise and call ahead. And don't forget your binoculars.

THE BIG EVENT	Place & Time	Why They Do It

Polar Bears wander the frozen wilds of Canada, where they feed on seals. Come summer, when their access to seals is limited, these giants and their cubs are forced south, toward the shores of Hudson Bay.

Wapusk National Park, near Churchill, Manitoba, mid-Oct. to early Nov.

As the ice melts and is pushed south, the bears are deposited on the wind-swept tundra around Hudson Bay, where they wait for the bay waters to freeze again. This is one of the largest viewable concentrations of polar bears in the world.

Monarch Butterflies bejewel forests in southern Ontario throughout summer. In fall, great swarms flee its icy fingers for Texas and Mexico, where they perch for hours. The trees are aflame with their delicate orange wings.

Kickapoo Caverns State Natural Area, Tex., late Oct. Other sites: Point Pelee National Park, Ontario, Sept.; Cape May Bird Observatory, N.J., late Sept. to early Oct.

Where milkweed blooms, monarchs follow, for it's what they eat and where they lay eggs. No single butterfly makes the round-trip. Several generations are born en route, so only the offspring complete the journey.

Humpback Whales born in the North Atlantic breed near the West Indies and summer along the east coast of the U.S. The seas off Cape Cod are a favorite feeding ground. Sometimes they can be seen breaching, their bodies hurled skyward by mighty flukes.

Stellwagen Bank Marine Sanctuary, Mass., April to Oct. Other sites: Glacier Bay National Park, Alaska, June to early Aug.; Hawaiian Islands Humpback Whales National Marine Sanctuary, Maui, Nov. to March.

Stellwagen Bank, a rich source of small fish and crustaceans, is a prime site for humpbacks to display a diverse repertoire of hunting skills. To corral prey, they may swim together in synchronized groups or blow bubble curtains around them.

Sandhill Cranes flying in spring to their nesting grounds stop off at Nebraska's Platte River to refuel along the way. This is when 40,000 silhouettes may fill the broad river basin at dusk. At dawn, they rise again like one massive winged veil.

Lillian Annette Rowe Sanctuary and Crane Meadows Nature Center, Neb., March to mid-April. Other sites: Bosque del Apache National Wildlife Refuge, N. Mex., mid-Nov. to mid-Feb.; Monte Vista National Wildlife Refuge, Colo., spring and fall.

The Platte River is a staging ground for northern migration. All told, a half-million cranes lay over to fatten themselves on snails, worms, and nearby cornfields. Each bird packs about a pound of fat onto its nimble frame in an average 29-day stay.

Caribou winter in small groups throughout the mountain valleys of Alaska and Canada. With the coming of summer, the Porcupine herd crosses the Brooks Range for the coastal plains, where hundreds of adults and newborn calves graze on a lush carpet of flowering herb plants.

Arctic National Wildlife Refuge, Alaska, June to Aug. Other site: Mulchatna River, Alaska, June through Aug.

Arctic NWR plains are isolated from wolves, grizzlies, and other predators. They also yield a bounty of highly nutritious vegetation, like sedge, willow, and herbs. So the Porcupine herd, at 170,000 strong, journeys here to bear young and fatten up for winter.

Horseshoe Crabs emerge from the sea each spring to lay billions of eggs and ensure their genetic future. Millions of shorebirds around the Delaware Bay region feed on the eggs, fattening themselves for their own migration north.

Reeds Beach, Cape May, N.J., May 10 to June 10. Other site: Bombay Hook National Wildlife Refuge, Del., May 10 to June 10.

Female crabs leave Delaware Bay's muddy floor to lay some 80,000 eggs each. This pattern, which predates the dinosaurs, sustains the world's second largest concentration of migrant shorebirds, including red knots, dunlins, and ruddy turnstones.

THE BIG EVENT	Place & Time	Why They Do It
Raptors soar throughout the continent. During fall migration, they converge along North American flyways. Sixteen species have been spotted at Hawk Mountain, where eagles skim the ridges and kettles and broad-winged hawks circle in updrafts.	Hawk Mountain Sanctuary, Pa., Sept. through Nov. Other sites: Cape May Point State Park, N.J., Sept. through Nov.; Golden Gate National Recreation Area, Calif., late Sept. through Oct.; Grand Canyon National Park, Ariz., mid-Sept. to mid-Oct.	Like most birds, raptors migrate north to south, some staying within the continent, others heading as far south as Argentina. They can cruise for hundreds of miles largely by riding currents deflected off hills and ridges, and thermals made of rising columns of air.
Bison once symbolized the American West, ranging the grasslands in great rambling herds. Now only 150,000 remain. But they are no less mighty during rut season at the National Bison Range. Bulls gather to do bloody battle, banging their massive hairy heads.	National Bison Range, Mont., July and Aug. Other site: Tallgrass Prairie Preserve, Okla., July and Aug.	Bulls live like bachelors most of the year. During rut season, they assemble to win mates and breed with cows. Once done, they return to their bachelor existence. The cows bear one calf each in spring and remain in small groups throughout the year.
Loggerhead Sea Turtles live in all but the coldest oceans, but nest only on subtropical and southern temperate beaches. In Florida, under a moonlit sky, they waddle from the sea, laden with hundreds of eggs for deposit in soft sand hollows.	Sebastian Inlet State Recreation Area, Fla., June through Aug.; Canaveral National Seashore, Fla., June and July.	Central Florida has the largest nesting population in the Western Hemisphere. After the eggs incubate for two months, the hatchlings erupt en masse and scurry for the sea, where they live until they are ready to return and bury their genetic treasure.
Grizzly Bears spend about half their lives in dens. In summer, North America's largest land predators gather along salmon streams. Then more than 30 mighty bruins congregate near McNeil River's cascades, poised to feast on pink-bellied fish.	McNeil River State Game Sanctuary, Alaska, June through Aug. Other sites in Alaska: Brooks Falls, Katmai National Park, June and July; Tongass National Forest, June and July; Kodiak Island National Wildlife Refuge, June through Aug.	In summer, grizzlies store away enough fat and protein to survive the winter, sometimes doubling their size. To build such bulk, they rely on salmon to return from the Pacific Ocean to their natal streams.
Mexican Free-tailed Bats migrate by the millions to stateside hollows, like Carlsbad Caverns. In summer, they whirl up from the caves in great tornadoes to feed on night-flying insects. At dawn, they dive-bomb back into the darkness.	Carlsbad Caverns National Park, N. Mex., July and Aug. Other sites: Eckert James River Bat Cave, Tex., July and Aug.; Bat Condo, University of Florida, Gainesville, July and Aug.	Crossing the border from central Mexico, female free-tailed bats inhabit southern maternity caves to bear their young, while adult males may fly farther north to feed on a bounty of moths. In fall, they reassemble and fly south together.
Elephant Seals may look fat and lazy, but their ambitious dual migration spans the Pacific Coast not once but twice. Come winter, they pause at San Miguel Island to bear pups and battle for mates. In summer, they return to shed their massive skins.	San Miguel Island, Channel Islands National Park, Calif., Dec. to Feb. and April to Aug.; Año Nuevo State Reserve, Calif., Dec. to Feb. and April to Aug.	The double round-trip, between Alaska and California, is unique among vertebrates. At 12,500 miles, it is also the second longest mammalian migration. Over 20,000 seals, each weighing up to 6,000 pounds, wallow on San Miguel's wide-berth beaches.

Wildlife Destinations

◆

Throughout North America, parks, sanctuaries, and refuges from Florida to Alaska, Maine to Baja California, offer incomparable wildlife watching. To get close to wild animals, you can take off on hiking trails, float rivers in a canoe, or scuba dive and snorkel through undersea worlds.

Baxter
State Park
Maine

C H A P T E R 4

The early reports are in, and the news isn't good. The obituaries have declared that big sky and open spaces on America's East Coast are as extinct as the auk, killed off by strip malls, claptrappy development, and the national mania for wide and fast roads. Obviously, those issuing the reports have never been to Maine's **Baxter State Park**, a half-day's drive north of Boston, well beyond the clutter of the hungry suburbs, and even well beyond the vanishing Robert Frost New England of church spires and valley farms. To put the early reports to rest, all one need do is set off into Baxter's tangled wilderness. ◆ Let's say you choose to scramble up **South Branch Mountain** in the park's northeast corner. You grunt and sweat your way up through groves of birch and maple, stumbling over roots and rocks slick with moss and lichen. You gingerly step over moose scat, which look like quail eggs crafted from a rough brown papier-maché. (They're recognizable because enterprising Mainers have collected them, var-nished them, and made them into earrings and pendants to sell to tourists at roadside shops to the south.) ◆ At the ledgy summit, ringed with blueberry bushes and lorded over by red-tailed hawks soaring on updrafts, you drop your daypack, take a long pull of water, and finally look around. Slate-blue peaks rise distantly to the south, like dorsal fins surfacing through a dense sea of dark green. To the northwest, the landscape flattens and unfurls toward Canada, pocked by shimmery cerulean lakes left like pocket change by retreating glaciers some 12,000 years ago.

An expanse of forested wilderness, cloaked in an awesome silence, provides a home for moose and other creatures of the North Woods.

Bull moose pair up with cows and mate after vying for dominance in fall.

Pacific tree frogs (preceding pages), small amphibians that stay close to water, are the most commonly heard frogs in Pacific Coast woodlands.

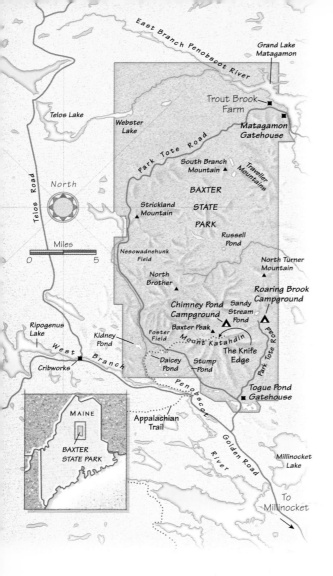

industrial timberlands and wild lakes laced with swaths of unbroken forest. They travel via dusty logging roads through timber company checkpoints and tumble down rushing rivers by canoe and raft. Mostly they're in the company of moose, white-tailed deer, eastern coyote, bald eagle, and black bear. Meeting up with a logger, fisherman, hiker, or canoeist is invariably cause for a chat.

Wildlife Comes First

The crown jewel of Maine's North Woods is **Baxter State Park**, home to **Mount Katahdin**, the state's highest and most rugged mountain. Roughly the same size as Voyageurs National Park in Minnesota, Baxter is 204,733 acres of mountains, woods, and rivers off-limits forever to development. It has little in common with state parks more commonly found in the East, where the megafauna endure blaring radios and melting popsicles and tend to be less than charismatic.

Visitors to Baxter are put on notice that this isn't their average state park when they enter the gate and are handed a copy of park rules. A brief synopsis: No radios. No vehicles more than nine feet tall (ergo, no RVs). No motorcycles. No motorboats. No TVs. No ATVs. No generators. No pets. No cell phones. Haul out all your own trash. Don't expect showers or even running water. In short, rule No. 1 is this: the park is managed for wildlife, not for the whim or entertainment of visitors.

The park's iconoclastic management can trace itself back to Percival Baxter, the wealthy son of a Maine canned-foods magnate who wanted to give something back to his state. When he was governor, Baxter tried valiantly to convince the state legislature to

But what captures your attention most is space. Miles of it. Montana-sized space. Then you notice that this immense space is packed in a silence so massive that it has dimensions, a silence so vast that it makes its own rushing sound deep within your head. In that, it hasn't changed since Thoreau ventured here a century and a half ago and pronounced the silence "more impressive than any sound."

Though it's not widely known, the fact is that fully one-fourth of New England remains uninhabited nearly four centuries after Europeans first settled the region. When travelers enter Maine's **North Woods**, they journey into a complicated patchwork of

Racoons (left), easily identified by their distinctive black masks, feed at night near sources of water.

Common golden-eyes (opposite), year-round residents in the Northeast, raise their broods in the park's ponds and waterways.

White-tailed deer (below) are seen in meadows at the edge of woodlands, nibbling on grass, shrubs, and trees.

buy and protect the land surrounding Katahdin, the state's highest mountain. The legislature was unreceptive. Irked but undaunted, Baxter started buying up the land in 1925 using his own fortune, donating the land to the state as he went. For the remainder of his life, he snapped up bits and pieces as they became available, handing over the last parcel in 1962 when he was 83 years old. Among his stipulations: that the park remain "forever wild," and that no Maine resident be charged to visit. Baxter also left a handsome endowment to help fulfill his vision, and to this day virtually no state funds are used in managing the park.

Today, all management decisions are weighed against Baxter's wish to preserve the land's wild character. For that reason, camping is strictly limited, certain trails are closed when their capacities are reached, and the single 41-mile dirt road that joins the north gate and the south gate is lightly maintained at best – the most expedient way to enforce the 20-mile-per-hour speed limit and to discourage tourists who prefer to sight-see from the comfort of their cars.

Next to car touring, the least creative way to explore the park is practiced by all-too-many visitors: they come, they climb Katahdin, they leave. Granted, the summit of the nearly mile-high massif is dramatic – all craggy and medieval, carved by glaciers and blasted with winds. It's the northern terminus of the **Appalachian Trail** and home to the **Knife Edge**, a mile-long trail between the mountain's two peaks, famous for triggering vertigo in those who've never before had a fear of heights.

For wildlife seekers, Katahdin's summit serves better as a reference

point than as a destination. It's in the untrammeled valleys, along forested shores of lakes and ponds, and atop the park's 45 less daunting peaks that you'll find more abundant and varied wildlife. Visitors hoping to put the park in context before trending into the woods have good options. From **Millinocket Lake** near the south gate, you can hire a floatplane for a midaltitude tour of the park, buzzing around Katahdin, through mountain passes, and over the lake-specked forest. Experienced pilots can pick out the wildlife in this sprawling mosaic – an eagle soaring above, maybe, or moose swimming across a marsh-edged river below. But mostly what takes the breath away is the sense of expansiveness, a vast forest broken here and there by clear-cuts and logging roads, but with no sign of habitation.

Along the park's southern edge runs the **West Branch** of the **Penobscot River**, a historic, tumultuous artery of commerce. From the mid-19th century into the 1970s, the river's chaotic rapids were often clotted with pulpwood on its way to the paper mills of **Millinocket**. (The state finally banned log drives because of their harm to the riverbeds.) Today, the river is more often than not clotted with the rafts of whitewater outfitters, several of which are based in Millinocket. A day's journey offers a tour of the park's backyard, from a heart-pounding descent through a pinched and rocky gorge called the **Cribworks**, followed by intermittent drifting through the calm stretches, where passengers can kick back and admire Katahdin's dour peaks, scan for beaver and moose at the river's edge, and admire from close range the punkish tufts of the common mergansers.

"Moosey and Mossy"

"All moosey and mossy" is how Henry David Thoreau described the North Woods after a journey to Maine in 1846. He and three companions traveled by bateau up the Penobscot River, then struck out on foot for the summit of Katahdin. Two days later, he cleared timberline and came face to face with a nature that was utterly unfamiliar to him. With damp and furious winds, nature seemed possessed of an elemental fury profoundly unlike what he had come to know in the sylvan woods of Massachusetts.

Thoreau later wrote that Katahdin's heights were "an undone extremity of the globe," a piece of the earth as yet unfinished by God and thus wholly unfit for the vain and prying eyes of man. A bit spooked by it all, Thoreau turned back and sought shelter below treeline, although even here he felt ill at ease. "The trees are a standing night, and every fir and spruce which you fell is a plume plucked from night's raven wing."

Thoreau didn't actually reach the summit, but he is generally numbered among the first half-dozen people to scale Katahdin. Despite his profound unease with nature raw and elemental, he returned twice more to Maine to plumb the deep woods by canoe, in 1853 and 1857. And Katahdin? Well, he left the mountain alone.

Martens (above), members of the weasel family, are omnivores that hunt small mammals, birds, and insects and also consume nuts and berries.

Henry David Thoreau (left) was not prepared for the inhospitable conditions that he encountered on his attempt to ascend Mount Katahdin.

Dense forests (right) carpeted with moss give the park's remote areas a silent, peaceful character.

Stalking Moose

For most Baxter visitors, a trip is incomplete without coming upon at least one moose, the herbivore king of the park. While moose can sometimes be spotted from a car, the most exhilarating place to encounter one is in the back-country, and Baxter maintains 175 miles of hiking trails that lead you there. For moose, summers are for feasting, when nature lays out a boun-tiful smorgasbord that allows them to attain their full weight of more than 1,000 pounds (less for cows) and bulk up for the leaner months of fall and winter. When feed-ing in summer, moose typical-ly consume some 40 or 50 pounds of food every day, mostly in the early morning and evening. Their preferred meal? Plants rich in sodium, like pond lilies and the roots of other aquatic plants. If you see moose standing up to their bellies in a pond pawing at the bottom, they're digging up dinner. Come the long Maine winter, when lakes ice over for months on end, moose turn their attention to the new and tender shoots of fir trees, and in spring they dine on the sap-infused bark of ash and maple.

Baxter boasts a number of notable moose watering holes. Among the most reliable is **Sandy Stream Pond**, just a few minutes' walk from **Roaring Brook Campground** near Katahdin's eastern base. The pond is so scenic, with its mountain-ous backdrop, and moose so reliably seen here that visitors tend to speculate that it's some Disneyesque animatron placed by state tourism authorities. (The parking area often fills up at dawn with Katahdin hikers,

after which the road to Roaring Brook is closed off. Ask at the gate if a Sandy Stream Pond pass is available, which allows you three hours to explore the area before you need return the pass.)

In the park's southwest corner, you'll find other dependable spots to stalk moose in summer, including **Stump Pond** near **Abol Campground**, the lakeshore around **Kidney Pond**, and **Daicey Pond**, where travelers with sufficient foresight can rent rustic lake-

side cabins for the night. Come fall, moose take to the hoof, when bulls actively seek out mates. Bull moose first assemble in small knots of three or four, then compete in what amounts to shoving matches, pressing antler against antler. Status established, the dominant moose then prowls widely, listening for the love call (more like a love "blat") of lone cows. Rutting moose can lose as much as 20 percent of their body weight in their frantic search for mates, and human encounters with normally placid moose can be a bit friskier come autumn. (Leave ample room between you and them.) Promising areas for spotting late-season moose include the open areas around **Trout Brook Farm**, a few minutes' drive from the park's north gate, and **Nesowadnehunk Field**, on the park's western edge. Those willing to expend a bit more energy can hike seven miles to **Russell Pond** in the park's center, where moose are reliably spotted year-round.

Loons and Wolves Beckon

Some moose watchers become so intent on stalking their quarry that they fail to notice the park's other wildlife. Don't make that mistake. Sharp-eyed red foxes pounce on mice in the park's meadows and prowl around shrubby, sandy areas. Look at **Nesowadnehunk**, **Foster Field**, and **Trout Brook Farm**, and around the **Togue Ponds** at the park's south entrance. The sleek marten, beaver, mink, and otter frequent many of the park's waterways. Bobcat (or bobcat tracks) are spotted by a lucky few; lynx are known to prowl near the park, but it's unclear if they're often within its borders.

Bird life is abundant at Baxter. Loons can typically be sighted along the wild shores of the larger lakes, like the **Togue Ponds** and **Webster Lake** and **Matagamon Lake**. At the least, you'll hear their eerie cackle and laugh at night, echoing off the forest wall. Above the ridges, watch for bald eagles and hawks, including red-shouldered and red-tailed hawks. Also listen for the spirally, fluty trill of the hermit thrush, which imbues the forest with a Hobbit-like enchantment.

Coyotes fleetingly lope across a trail. The eastern coyote, distinct from its western cousins, is shrouded in some mystery, but most biologists believe it to be a new species that evolved in recent times to adapt to the changing ecology of New England forests.

Designed By Committee

Moose weigh up to 1,400 pounds and can be somewhat intimidating when suddenly encountered along a backcountry trail. Despite their impressive bulk, though, there's something fundamentally cartoonish about them, which usually makes an encounter more entertaining than terrifying. The long spindly legs. The barrel-shaped body covered with bristly hairs. The small bulging eyes. The scraggly beard. Their comical loping escape as they crash through the brush with a notable lack of finesse. And that bulbous snout! Thoreau wasn't far from the mark when he said, "The moose is singularly grotesque and awkward to look at."

Moose belong to the deer family – their name comes from the Algonquin word for "twig-eater" – and while it may appear that they were designed by a distracted committee, their unique characteristics actually mark a number of traits handy for surviving in these dense northerly woods. The location of their bulging eyes gives them almost 360-degree vision (although not very good vision – they depend more on hearing and smell). Moose may be gangly when they try to gallop away, but those rangy legs allow them to browse gracefully through riverside marshes (their front legs are actually longer than their rear, the better to scramble over low obstacles). The thick, bristly fur is hollow, which makes moose more buoyant for swimming and keeps them warmer in winter. And the big rabbit-like ears aid their extraordinarily acute hearing, able to detect the footsteps of an approaching animal as many as three minutes before a human would hear them.

The one mystery is that silly beard, technically called a "bell." Biologists haven't yet figured out what that's all about.

Lynx (opposite, above) prowl the area around the park, hunting their favorite food, snowshoe hares.

Common loons (opposite, below) emit a haunting repertoire of wails and yodels. They are commonly heard at night in summer.

Moose (below) lose the velvet that covers their antlers by late fall. The antlers themselves are shed in winter.

The coyotes are suspected to have emerged from a hybrid of Minnesota coyotes and timber wolves. They're slightly larger than western coyotes and marked by a straight tail with a black tip. Keep an ear tuned for their howl in the night.

And then there are the black bears, of which the park has a population variously estimated at 160 to 200. Your best bet for bears – whether to find or avoid them – is August in the **Traveler Mountains** in the northeast corner of the park. The bears migrate here to gorge on the plentiful berries that thrive along the ledges in season. The park has done a stellar job keeping visitors from feeding bears, and as such they'll be wild, shy, and quick to disappear if something gives them a start. If you fail to spot a bear on your excursion, you're all but certain to see the unusual bear scat on the trails: iridescent rolls of partially digested blueberries that look rather like a manufactured snack from a convenience store.

A week at Baxter is by no means too much. Explore the valleys and remote ponds at the outset, then work your way up the far-flung peaks. Each day you'll catch glimpses of Katahdin's angled peak rising with a hulking sternness from the forest – sometimes stark against an azure sky, other times capped with slick lenticular clouds. You may find yourself drawn to an ascent by the end of your sojourn. By all means, heed the call. From the soaring summit, you can embrace an endless space, one that will now seem all the less empty for your explorations.

TRAVEL TIPS

DETAILS

When to Go

Summer is the best time to visit northern Maine. Spring is short, wet, and blustery, and winter clamps down early and refuses to budge. Summer temperatures are usually in the 70s, with cool nights and brief spells of heat and humidity. The bugs are often at their worst in early summer, so be prepared with insect repellent.

How to Get There

The closest airport is Bangor International, about 86 miles south of the park. Portland International is about 220 miles away.

Getting Around

There are only three ways to get around the park: by car, foot, and bicycle (on maintained roads only). Motorcycles are prohibited, and size restrictions exclude many recreational vehicles. There is no road access in winter.

Backcountry Travel

Day-use backpacking requires no permit. Camping is allowed only in designated areas. Winter camping (December 1 to April 1) is limited to parties of four with appropriate backcountry experience, and must be arranged at least two weeks in advance.

Handicapped Access

Facilities are limited, with the exception of some tent sites and bathrooms.

INFORMATION

Baxter State Park

64 Balsam Drive; Millinocket, ME 04462; tel: 207-723-5140.

Katahdin Area Chamber of Commerce

1029 Central Street; Millinocket, ME 04462; tel: 207-723-4443.

Maine Tourism

P.O. Box 2300; Hallowell, ME 04347; tel: 800-533-9595 or 207-623-0363.

CAMPING

Campsites must be reserved by mail or in person. The best sites fill up fast, so make reservations well in advance. January isn't too early to make reservations for a summer trip. Camping is permitted from May 15 to October 15. For information, call the Reservation Clerk at 207-723-5140. For information on camping outside the park, call North Maine Woods Inc., 207-435-6213.

LODGING

PRICE GUIDE – double occupancy

$ = up to $49 $$ = $50-$99
$$$ = $100-$149 $$$$ = $150+

Big Moose Inn

P.O. Box 98; Millinocket, ME 04462; tel: 207-723-8391.

This cedar-shake inn is set between two lakes just eight miles from the park entrance. Eleven old-fashioned rooms are furnished with oak and pine antiques. They share three baths and vary in size from single to family units. Twelve rustic log cabins, some lakeside, feature kitchenettes and private baths. Amenities include a restaurant, canoe rental, and two campgrounds. $

Chesunook Lake House & Cabins

Route 76, Box 656; Greenville, ME 04441; tel: 207-745-5330.

Accessible only by boat or seaplane, this peninsula hideaway is an oasis of domesticity about 30 miles south of the park. The three-story farmhouse was built in 1864 and is situated on Chesunook Lake, with views of Mount Katahdin. Original pressed-tin walls and ceilings set the tone inside, where sunny guest rooms overlook the lake. Rustic housekeeping cottages accommodate 12 to 15 people. Canoe and motorboat rentals are available. $$$$

Frost Pond Camps

Box 620, Star Route 76; Greenville, ME 04441; tel: 207-695-2821.

Eight housekeeping cabins and 10 campsites are set on Frost Pond about an hour's drive from Baxter State Park. Mount Katahdin is visible from the pond, where bald eagles and ospreys pay regular visits. Cabins sleep two to eight and are equipped with gas lights, refrigerators, stoves and ovens, and pit toilets. One large cabin has running water; a coin-operated shower services the others. Boat rental is available. $

Katahdin Shadows

P.O. Box HB; Medway, ME 04460; tel: 800-794-5267 or 207-746-9349.

The complex includes a motel, five cabins, and a campground about 25 miles from the park. The motel's 10 comfortable rooms vary in size. Heated log cabins accommodate up to 10 people. Bathroom facilities are just a few steps away in the main lodge. The motel features a hot tub and outdoor heated pool. Boat rentals are available. $

Metagamon Wilderness Campground

P.O. Box 220; Patten, ME 04765; tel: 207-528-2448.

These housekeeping camps are set on the East Branch of the Penobscot River near the park's north entrance. Cabins sleep six to nine people. Bathrooms are located in a central shower facility. Boat rentals are available. $

Mount Chase Lodge

Upper Shin Pond; Patten, ME 04765; tel: 207-528-2183.

On Shin Pond, 15 miles from the park entrance, this complex of log buildings has eight guest rooms and five cabins. Large rooms in the main house have private baths, as do the heated cabins, which sleep up to eight. Amenities include a waterfront lounge, dining room, and guide service. $-$$

Northern Pride Lodge

HC 76, Box 588; Kokadjo, ME 04441; tel: 207-695-2890.

On scenic First Roach Pond about 40 miles from the park, this lodge and campground are surrounded by thousands of acres of private-ly owned timberland. The house, built in 1896, has large rooms and lake views. Expect daily moose sightings from May to September. Guests will enjoy a hearty Maine woods breakfast. Lunch and dinner are offered at extra cost. Boat rentals are available. $$

TOURS & OUTFITTERS

Allagash Wilderness Outfitters

Box 620, Star Route 76; Greenville, ME 04441; tel: 207-695-2821.

Complete outfitting for two- to seven-day canoe trips on area rivers and lakes.

Katahdin Air Service

P.O. Box 171; Millinocket, ME 04462; tel: 207-723-8378.

Ten flight packages from scenic 15-minute and one-hour flight-seeing tours to all-day drop-offs at secluded lakes and lodges.

Magic Falls Rafting Company

P.O. Box 9; West Forks, ME 04985; tel: 800-207-7238.

One- and two-day guided raft trips on the Kennebec, Penobscot, and Dead Rivers.

North Country Rivers

P.O. Box 47; East Vassalboro, ME 04935; tel: 800-348-8871.

Rafting trips on area rivers, with lodging or camping.

Excursions

Acadia National Park

P.O. Box 177; Bar Harbor, ME 04609; tel: 207-288-3338.

One of the 10 smallest national parks (40,000 acres) in the United States, Acadia is also one of the nation's jewels, a place where rugged mountains and forests press to the sea, and where a day's hike can take you from an alpine meadow to a rocky beach. The park also features the only fjord on the Atlantic Coast. Readily viewed wildlife include white-tailed deer, beavers, red foxes, sea urchins, porpoises, bald eagles, herring gulls, and humpback, finback, and minke whales.

Allagash Wilderness Waterway

Bureau of Parks and Recreation; Maine Department of Conservation; State House Station 22; Augusta, ME 04333; tel: 207-287-3821.

Established in 1970, this 92-mile lake and river corridor offers some of the best canoeing in the region. Although the river flows through commercial timberland, a 500-foot buffer zone protects trees from harvesting. Bald eagles, loons, and moose are commonly spotted along the lakeshores and riverbanks. Observant hikers may also find fossil-bearing rocks. Most canoeists take a week or more to float the entire route. Several portages are necessary. Dozens of campgrounds are available along the way.

Green Mountain National Forest

231 North Main Street; Rutland, VT 05701; tel: 802-747-6700.

You'll find some of New England's best hiking and mountain biking in this mountainous, 350,000-acre national forest. Black bears and moose abound. The former are most visible in autumn, when they comb the woodlands for acorns and beechnuts. Moose have increased almost to their limit of sustainability. The forest has one of the region's richest populations of river otters. Also abundant are beaver and mink, as well as many species of birds, including the rare northern goshawk.

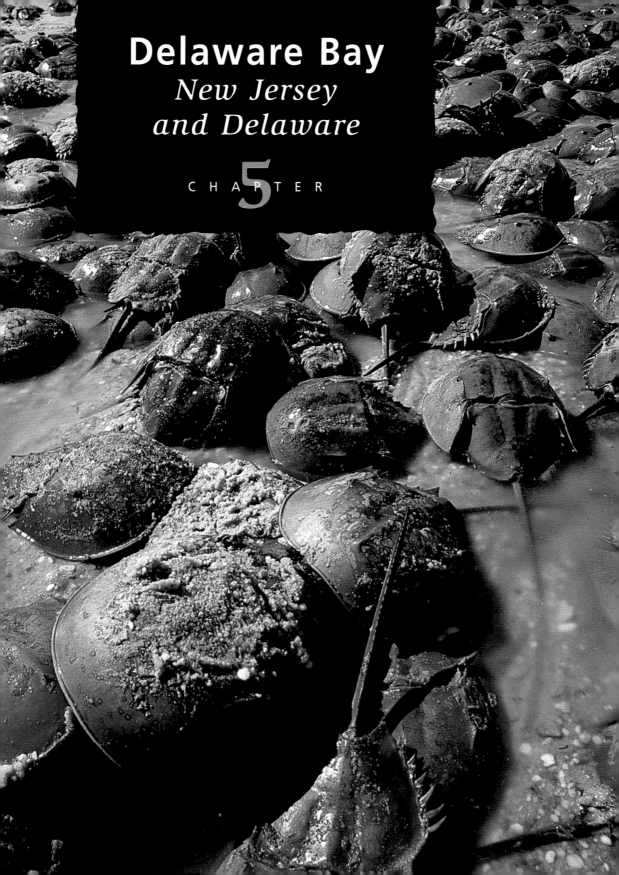

Delaware Bay
New Jersey and Delaware

CHAPTER **5**

Driving a cool, blue wedge between **Delaware** and **New Jersey**, rimmed by beaches and marshes whose vastness defeats human eyes, is a bay out of place and time. Humpback and finback whales forage here. Flocks of snow geese frost winter marshes. River otter abound, migrating monarch butterflies cluster, and in May, when full-moon tides ferry the hosts of horseshoe crabs onto bayshore beaches, hundreds of thousands of shorebirds gather to the bounty of their eggs – a spectacle of global significance. ◆ Bracketed by metropolitan centers, Delaware Bay lies within a three-hour drive of 45 million people. Yet it retains much the same character that Captain Samuel Argall found in 1641 when he and his crew first navigated the waters of the "Wihittuck" – literally, the "River of the Lenape [Indians]" – which he renamed in honor of his excellency Thomas West, Baron De La Warr. ◆ You have much to discover here, too, and at any season.

A bay for all seasons, this is the place to see the spectacle of migrating birds, butterflies, and marine animals.

Few places on this planet offer more natural spectacles more often than Delaware Bay. Perhaps none. ◆ Winters are kind to the region, prolonged freezes rare. Water, even fresh water, remains open and free of ice most years. Between November and March, thousands of hardy black ducks and elegant pintails gather in the upriver reaches of New Jersey's **Maurice River**, a federally designated Wild and Scenic River that is best known for its stands of sensitive joint vetch, a globally endangered plant, and autumn concentrations of soras, rather secretive marsh birds. ◆ At Delaware's **Bombay Hook** and **Little Creek National Wildlife Refuges** as well as the marshes near **Heislerville**, New Jersey, tens of thousands of snow geese

Horseshoe crabs crawl ashore in May and deposit their eggs, offering a feast for migrating shorebirds.

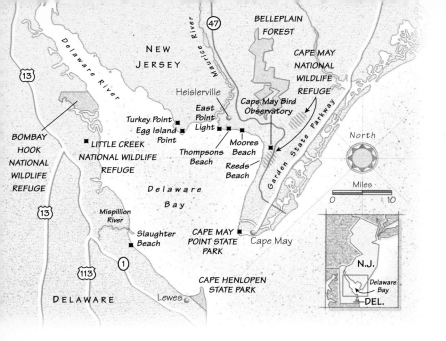

probably no place on the planet that boasts more great horned owls than the marshes of Delaware Bay. At **Turkey Point**, a wooded neck reaching out toward **Egg Island Point**, the owls sally forth each evening, perching prominently on the edge of the marsh. On still nights three, four, even five pairs may lie within earshot – a chilling sound for cottontail rabbits; music to the ears of would-be owl spotters.

Spring migration begins in earnest in March. **Cape Henlopen**, Delaware, is a mini-migrant trap and a place to watch coastal hawk migration. By late April, many of the region's breeding birds have returned, and come May waves of warblers are passing en route to Canadian Zone woodlands.

forage. Rooting in the marshes for spartina grass roots, they raise muck-stained faces to plumb the world for danger. The clamor of the flocks rises and falls as eagles maneuver across the sky. Suddenly one of the great winged hunting birds turns. Wedding the strength of its wings to gravity's pull, it falls. The flocks thunder aloft. Sometimes the eagle wins and claims a bird from the flock. More often, the geese elude the charge. Predator versus prey – business as usual on the marshes of Delaware Bay.

By February, with great horned owls incubating eggs, and March, with spring peepers making the air ring with amphibian voices, spring reigns supreme. There is

The Great Crab Feed

But the region's most famous natural attraction is the great juxtaposition of life that occurs along the beaches in May. There, on every rising tide, hundreds of thousands of horseshoe crabs clamber ashore. Male suitors in tow, the hubcap-shaped female crabs excavate nests and deposit the tiny gray-green eggs that feed multitudes of migrating shorebirds.

Thousands of ruddy turnstones, sanderlings, and red knots gather to the feast, larding on the layers of fat they will need to fuel their last nonstop flight to arctic breeding grounds. Thousands of visitors head for the mouth of the **Mispillion River** in Delaware or **Reeds** and **Thompsons Beaches** in New Jersey to witness the ribbon of life that fringes Delaware Bay between May 10 and June 10.

Come summer, the near-shore waters of the bay are alive with bottlenose dolphins. The region is a major birthing area for these winsome marine mammals, and bathers

delight in their proximity. Farther off shore, it is increasingly common to find humpback and finback whales. In fact, whale-watching boats put out of both **Cape May** and **Henlopen** from May to October. There are times when a whale surfaces and blows so close to a boat that passengers can feel its exhaled breath.

Hawks and Butterflies in Migration

Summer is all too short on Delaware Bay. The last northbound birds have hardly passed before the first southbound shorebirds reach the region's airspace, sometimes arriving before June 25. By the end of July, the migration peaks for some shorebird species, most notably short-billed dowitchers and lesser yellowlegs. In August, fall migration is in full swing. Guarding the eastern flank of Delaware Bay is **Cape May**, a place whose name is synonymous with migration.

Millions of songbirds navigate the peninsula's length, caught in a geographic funnel between ocean and bay. On the heels of every cold front, they crowd into woodlands and meadows at the peninsula's tip – warblers, vireos, thrushes, and other species bound for the tropics.

Massed concentrations of migrants, which are called fallouts, occur only a few times a season, but any time between August and November is good to see birds in Cape May, especially birds of prey. Between September and November, an average of 65,000 migrating raptors representing 15 species are counted at the

hawk watch platform at **Cape May Point State Park**. More than 1,800 peregrine falcons have been counted in a season; 21,000 hawks in a single day!

While the heads of hawk watchers crane skyward, another migratory phenomenon is moving at their feet. In September and October, monarch butterflies are on their way to Mexico. Ferried to the coast on northwest winds, thousands of these orange-winged insects pulse down beaches and gather at evening roosts. In years when monarch numbers are high, the branches of favored roost trees may bow beneath the weight of the insects. When morning sunlight finds them, the branches bloom with fiery orange blossoms.

These are just some of the many wonders that await you when you visit Delaware Bay. There are others, so many that they defy the region to hold them. You will simply have to come yourself and find your own fitting last words for this natural treasure trove.

Gull and shorebird species (left) compete for a meal at Reeds Beach on Delaware Bay.

Teams of birders (right) fan out over New Jersey to count birds and raise funds for conservation.

World Series of Birding

On the second Saturday of May, somewhere in the marshes of New Jersey, the air rings with the calls of birds and this familiar litany: "KING RAIL! Calling! Got it?" "Got it." "GOT IT!" "GOT IT!"

It's time for North America's most celebrated environmental treasure hunt, the World Series of Birding. Held every year since 1984 under the auspices of the New Jersey Audubon Society, the event has drawn up to 60 teams of top field birders who compete to find the most species of birds during a 24-hour period.

Teams are sponsored by corporations and conservation groups, and over half a million dollars is raised for bird conservation. Some of the teams spend months planning their strategy and days scouting routes that may exceed 500 miles.

The rules are simple. All birds must be found in New Jersey. All birds must be seen or heard by at least two team members. Every species counts as one. Winning teams have tallied 227 species. Over 270 species have been found by all teams – more than one-third of the birds found in North America.

TRAVEL TIPS

DETAILS

When to Go

Weather in the region is moderated by the ocean and bay. Winter and fall are relatively mild; spring is cool and breezy; and summer is warm and pleasant, with average temperatures in the 80s. Wildlife watching is best during the annual migrations. Horseshoe crabs come ashore to lay eggs in May and early June followed by thousands of ravenous migratory birds. Raptor migration peaks in September and October.

How to Get There

Atlantic City International Airport is 48 miles away from Cape May, Philadelphia International Airport is 87 miles. Car rentals are available at both airports. New Jersey Transit, 800-582-5946, operates daily bus service between Atlantic City and Cape May.

Getting Around

Cape Area Transit, 800-966-3758, operates a shuttle in the town of Cape May from April to October. High Roller taxi service, 609-884-5711, is available year-round.

INFORMATION

Chamber of Commerce of Greater Cape May

513 Washington Street Mall; P.O. Box 556; Cape May, NJ 08204; tel: 609-884-5508.

Lewes Chamber of Commerce

P.O. Box 1; Lewes, DE 19958; tel: 302-645-8073.

New Jersey Travel and Tourism

20 West State Street; Box 826; Trenton, NJ 08628; tel: 800-537-7397.

CAMPING

Cold Spring Campground

541 New England Road; Cape May, NJ 08204; tel: 800-772-8530 or 609-884-8717.

Wooded campground open from May to mid-October.

LODGING

PRICE GUIDE – double occupancy

$ = up to $49	$$ = $50-$99
$$$ = $100-$149	$$$$ = $150+

Henry Ludlam Inn,

1336 Route 47; Woodbine, NJ 08270; tel: 609-861-5847.

A favorite among birders, this 18th-century house is adjacent to Ludlam's Pond, less than 30 minutes from Cape May. The inn has five cozy guest rooms, each with private bath, three with fireplace. Period antiques and pegged floorboards secure a fine Colonial ambiance. Watch birds from the lakeside gazebo or from one of the inn's canoes. Indoors, sit down with a book from the birding library or browse through the inn's birding paraphernalia. $$-$$$

Mainstay Inn

635 Columbia Avenue; Cape May, NJ 08204; tel: 609-884-8690.

In 1872, two gamblers hired a famous architect to design a gentlemen's club. The result was a lavish Italianate villa, now part of the Mainstay Inn. Outstanding are the building's 14-foot ceilings, walnut furnishings, wrap-around veranda, and cupola. Two blocks from the Cape May beach and boardwalk, the inn has three buildings and offers an assortment of guest rooms, all with private bath. Amenities include afternoon tea, a library, flower garden, croquet, and swings. $$-$$$$

Manor House

612 Hughes Street; Cape May, NJ 08204; tel: 609-884-4710.

Guests at this Victorian inn enjoy exquisite breakfasts – fresh-squeezed juices and made-from-scratch gourmet entrees. Relaxation may take the form of a high-backed chair in the garden or a tome in the reading room. Some of the inn's 10 guest rooms overlook the garden, others the ocean; each has a private bath. $$-$$$$

New Devon Inn

P.O. Box 516; Lewes, DE 19958; tel: 800-824-8754 or 302-645-6466.

Situated in the center of the Lewes Historic District, this inn offers 24 guest rooms and two suites, each with private bath. The brick three-story hotel was built in the late 1920s and restored in 1989. Antique-filled rooms are individually decorated and stocked with fine linens and fluffy towels. Art Deco tones grace the lobby, which includes a sitting area and small music room. Amenities include six retail shops and a restaurant. $$-$$$$

Queen Victoria

102 Ocean Street; Cape May, NJ 08204; tel: 609-884-8702.

This Victorian bed and breakfast is known for its slavish attention to detail and strong emphasis on comfort, with such agreeable touches as flower gardens, afternoon tea, handmade quilts, and priceless antiques. Made up of three buildings, just a block from the beach, Queen Victoria has guest rooms of various size, all with private bath, some with whirlpool tubs and gas fireplaces. $$-$$$$

Wooden Rabbit Inn

609 Hughes Street; Cape May, NJ 08204; tel: 609-884-7293.

Nestled in the heart of historic Cape May, two blocks from several noteworthy restaurants, the Wooden Rabbit is small in stature but big in sentiment. The house was built in 1838 and is set on a street lined with gas lamps. Horse-drawn carriages pass frequently. Guest rooms are decorated with country prints and furnished with antiques and wicker. All rooms have private baths, king- or queen-sized beds, and double sofa beds. Suites offer an additional sleeping room. Guests are welcome to use the garden, library, beach chairs, and beach tags. $$$-$$$$

TOURS & OUTFITTERS

Cape May Bird Observatory

600 Route 47 North; Cape May Court House, NJ 08210; tel: 609-861-0700.

In addition to providing visitors with information about local bird and butterfly observation, the center offers workshops, nature weekends, regularly scheduled walks, and a host of seasonal programs and field trips.

Cape May Whale Watch and Research Center

Route 109 and Ocean Drive; Cape May, NJ 08204; tel: 888-531-0055 or 609-898-0055.

Cruises aboard a 75-foot motorized catamaran offer a narrated tour around the cape. Whales, dolphins, and pelagic birds are regularly encountered. April through November.

Cape May Whale Watcher

Miss Chriss Marina; 2nd and Wilson Drive; Cape May, NJ 08204; tel: 800-786-5445.

Whale and dolphin watching, birding, and dinner cruises are offered on the largest whale-watching boat in southern New Jersey. Sightings are guaranteed or passengers ride again free.

Wildlife Unlimited

10 Wahl Avenue; Dias Creek, NJ 08210; tel: 609-884-3100.

Educational boat tours through the coastal salt marsh, one of the world's most productive ecosystems. Sights may include heron rookeries, nesting ospreys, and a huge colony of laughing gulls. Dolphin sightings are not uncommon.

Excursions

Chincoteague National Wildlife Refuge

P.O. Box 62; Chincoteague, VA 23336; tel: 804-336-6122.

Famed for its wild ponies, this 13,682-acre coastal refuge is also home to sika and white-tailed deer, endangered Delmarva Peninsula fox squirrels, otters, and foxes. Three-quarters of the ponies are owned by the Chincoteague Volunteer Fire Department, which rounds them up and steers them across the inlet each year for an auction of excess foals. Three hundred species of birds may be seen here at various times of the year. From fall to early spring, Chincoteague's marshes are visited by migrating waterfowl, including vast flocks of snow geese, mallards, pintails, and black ducks.

Delaware Water Gap National Recreation Area

River Road; Bushkill, PA 18324; tel: 717-588-2451.

"The Gap" takes its name from a dramatic notch cut by the Delaware River into the Kittatinny Mountains. The river flows for about 37 miles through the 70,000-acre recreation area. Along the way, it captures scores of tributaries, many of which cascade down shady hemlock ravines in thunderous veils of water. Hikers on the area's 60 trails – including a section of the Appalachian Trail – may see white-tailed deer, foxes, black bears, coyotes, bald eagles, and ospreys.

Wharton State Forest

R.D. 9 Batsto; Hammonton, NJ 08037; tel: 609-561-0024.

The forest's 110,000 acres lie in the heart of the Pine Barrens, a vast region of pine stands, streams, and swamps. It is home to coyotes, beavers, river otters, both red and gray foxes, white-tailed deer, bald eagles, sharp-shinned and red-tailed hawks, and pine warblers. Nearly 500 miles of sand roads, great for hiking, lead to the forest's many streams, marshes, and canoe-worthy rivers. Rare orchids and the bog asphodel, found only in New Jersey, grow in the pine, oak, and cedar forest.

Great Smoky Mountains
National Park
N. Carolina/Tennessee

CHAPTER **6**

et's face it: A wildlife safari at Great Smoky Mountains National Park in Tennessee and North Carolina is not at all like one on the plains of Kenya. Sure, a four-wheel-drive safari (a.k.a. sports utility) vehicle might come in handy on some of the dirt back roads. Sure, a safari jacket might be nice for carrying rolls of film, a jackknife, snacks, and insect repellent. Sure, the Smokies are home to a great variety of wild animals. But where are they? You can't see the animals for the trees. ◆ Under a smoky-blue haze, the seemingly endless mountains and ridges are covered with forests of deciduous and conifer trees and thickets of rhododendron, mountain laurel, and grapevines. This dense vegetation is fed by abundant rainfall, up to 80 inches annually on the peaks and 50 inches in the valleys. The waters seep into the soil and rush down the steep moun-tainsides, loudly cascading from one enormous, moss-covered boulder to another. Most of the park is a lush, deep-green jungle, an Appa-lachian jungle cloaking an astonishing panoply of wildlife. ◆ Few places in North America can match the

Lush forests harbor a bounty of wildlife, from the black bear to the greatest variety of salamanders in the world.

Smokies' wildlife statistics: 238 kinds of birds, 70 species of mammals, 60 kinds of fish, 29 types of salamanders, 23 kinds of snakes, 13 species of toads and frogs. With elevations ranging from 840 to 6,643 feet, the Great Smoky Mountains are host to plants and animals that might be found any-where from Georgia to Maine. Most of the animals are small and remain hidden as they go about their daily lives in the woodlands and streams. But with a little patience and some hiking, visitors can see plenty of wildlife in all seasons except winter and quickly understand why the park has been

Black bears are generally retiring, but they sometimes surprise visitors by making an unexpected appearance.

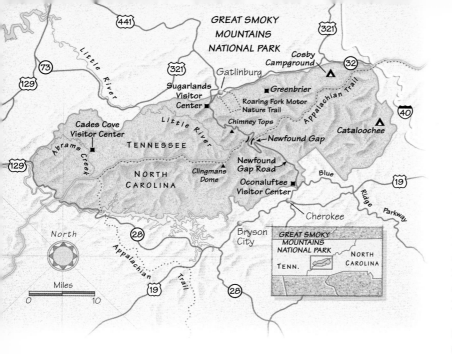

or trail. Though 400 to 500 black bears live in the park, 95 percent of them remain out of sight. Nonetheless, park rangers suggest talking, singing, or making noise in some way to avoid surprising bears that might be feeding in the woods and thickets on berries, acorns, or hickory nuts. Keep your distance if you see cubs, for a protective mother is probably nearby and won't care how cute you think her kids are.

Those wishing to do their bear watching from a vehicle should try the **Roaring Fork Motor Nature Trail**, which starts off Airport Road in Gatlinburg, Tennessee; the **Newfound Gap Road**, running across the mountains from Tennessee into North Carolina; and **Cades Cove**, in the western end of the park. Bears also have been seen often in the **Cosby Campground** and the **Greenbrier and Chimneys picnic areas**, but no bear-sighting guarantees are issued with these suggestions. Keep in mind, of course, that bears go into semihibernation from December to March or even May.

recognized internationally as both a Biosphere Reserve and a World Heritage Site.

The Elusive Black Bear

The black bear is by far the biggest attraction in size and visitor appeal. Standing six feet tall on its hind legs and weighing between 250 and 500 pounds, a mature bear instills both wonder and alarm in even the most jaded of backpackers. In an instant, a black bear can awaken our senses to their primeval core. The experience is unforgettable, something to tell the folks about back home for years to come. Bears, however, are not likely to appear around the next bend in the road

Meadows and Streams

Cades Cove, a flat area sheltered by the mountains, is one of the best places to observe not only bears but other wildlife. When the park was established in the 1930s, the government resettled the residents elsewhere and removed most of the buildings. Since then, nature has reclaimed the lands that had been farmed or logged since the late 1700s when Scotch-Irish, English, and German immigrants started moving into the mountains and displacing the Cherokee Indians.

Cades Cove is an exception. Here a few log and frame houses, a mill, and farm buildings from the 1800s are preserved for visitors to tour and learn about the folkways of the

southern Appalachians. Here, too, the meadows are mowed periodically. At dawn and dusk, groups of white-tailed deer graze on the grasses after spending the daylight hours resting in the surrounding foothills. Wild turkeys amble and strut about the open woodlands and adjacent fields as they forage for acorns, seeds, and insects. Woodchucks, known locally as groundhogs or whistle pigs, stand atop their burrows and scan the terrain for predators before venturing forth. Chipmunks scurry about gathering seeds and nuts for their underground winter pantries. At twilight, raccoons and gray and red foxes begin their nocturnal prowls for small prey.

After an absence of many years, river otters and beavers are back in the park. Last seen in 1936, otters were reintroduced in 1986 when park managers released 11 of them from North Carolina in **Abrams Creek** at Cades Cove. Since then, despite a few losses,

the otter population has grown, and some have moved to the **Little River** between Cades Cove and Sugarlands. They are bundles of energy, living, playing tag and sliding down riverbanks into the water as if they were hired entertainers.

Beavers apparently returned on their own in 1966 after disappearing from eastern Tennessee in the late 1800s because of heavy trapping. They tend to favor slow-moving streams, of which there are few in the Smokies, so their numbers probably will remain low. Look for their dams and chiseled trees along Abrams Creek. More skittish than the otter exhibitionists, beavers are likely to retreat to their underwater dens when humans approach.

A word to the wise: There are few roads in this the most heavily visited of all the national parks. From the spring flowering season through the fall foliage season, the park roads are jammed. So, plan to arrive

Northern cardinals (above), year-round avian residents, are among the nearly 250 species of birds seen and heard in the Smokies.

Flying squirrels (left) use their skin folds as gliders, enabling them to move quickly from tree to tree.

River otters (opposite), a native species successfully reintroduced into the park, are increasing their numbers and expanding their territory.

in Cades Cove at daybreak before the daily bumper-to-bumper caravan appears on the 11-mile, one-way loop road. Similar wildlife sightings can be had at **Cataloochee**, a cove on the North Carolina side of the park, but without the crowds. This old settlement is remote and visitor services are scarce. It is fairly close, however, to the 470-mile Blue Ridge Parkway, which runs through the Appalachians between Great Smoky Mountains National Park and Shenandoah National Park in Virginia.

Hit the Trail

For a full, genuine Smokies experience, getting off the road is of utmost importance. With 800 miles of trails, including 70 on the **Appalachian Trail**, you can find a variety of hiking experiences in the 521,621-acre park, ranging from gentle strolls to strenuous backpacking adventures. Even if a trail's destination is a tumbling waterfall, a spectacular scenic view, a virgin forest, flower-ing shrubs, or a patch of wildflowers, you will see wildlife. Ask rangers in the visitor centers at Sugarlands, Oconaluftee, and Cades Cove for recommendations to suit your physical capabilities and wildlife interests.

Bobcats, foxes, skunks, and many of the other animals are nocturnal, but some of them start moving about in the late afternoon. High in the mountains, in the spruce-fir forests of the **Clingmans Dome** area, northern flying squirrels leap, spread their legs apart to extend folds of skin on their undersides, and glide as much as 150 feet from tree to tree. They use their tails, legs, and "parachute" to steer and to control their speed. These aerial acrobats are active mostly at night, but occasionally one puts on a matinee performance. Listen for another squirrel, the red squirrel, which the mountaineers called the "boomer" because of its long, drawn-out, scolding sounds that can be heard over great distances. This small tree squirrel feeds on pine nuts, acorns, and hickory nuts, so it is seen, as well as heard, throughout most of the park.

Salamander Haven

The greatest variety and largest population of salamanders in the world except for the tropics make their home in the Great Smoky Mountains. Look under a rotten log or under a rock in a moist area and chances are a salamander will scoot away before you can figure out which of the 29 Smokies species it is.

Identifying salamanders is perplexing even if they stand still. The Appalachian woodlands, or Jordan's, salamander, for example, occurs throughout the southern mountains. It is dark gray or black. One form of this salamander, the red-cheeked, lives only in the Smokies and is distinctly marked, as its name indicates. Just to confuse things, another endemic species, the imitator salamander, has some individuals with red cheeks.

All salamanders are amphibians, but some species stay only on land, and some, including the red-cheeked, have no lungs. Salamanders discourage predators with skin secretions or by playing dead, but if bitten they can regenerate a tail or other body part. Don't bite one. And please, if you move a log or rock, put it back. It's a salamander's home.

The red eft (left), the juvenile form of the red-spotted newt, resides on land and, when grown, moves to the water.

The three-lined salamander (below) is found along slow-moving streams in the park's lowland habitats.

Indigo buntings (opposite, above) breed in the Smokies. The males define their territory with bursts of song.

Birding by Sight and Sound

Though the black bear is the Smokies' popular drawing card, it is the abundant bird life that attracts most wildlife enthusiasts to this national park. The number of resident and migrant bird species counted by bird-watchers, or perhaps more accurately by bird-listeners, is fast approaching 250. Because of all the topographical ups and downs and the dense vegetation, it is often easier to identify birds by sound than by sight. In the lower hardwood forests, listen for the red-eyed vireo, a canopy dweller that seems to be saying over and over again, *see me, here I am, up here, I'm here.* A staccato *teacher, teacher, teacher* indicates the whereabouts of an ovenbird, which builds a Dutch-oven-like nest on the ground. Birders claim that the *zeer, zeer, zeer, zee or zee-zee-zee-zoo-zee* of the black-throated green warbler is one of the easier warbler songs to detect. Like many of the park's winged migrants, it winters in Central America.

The birds vary with elevation, vegetation, and season. Species commonly found in the northeastern states, such as the black-capped chickadee and the red-breasted nuthatch, occupy the high elevations. Vivid specks of yellow and orange flickering high in the spruce and fir trees probably indicate the presence of Blackburnian warblers, one of the park's many warblers. A warbler seen more readily, the common yellowthroat, feeds on insects in shrubby areas along streams. The male's larger yellow patch extends from its throat to its breast.

In the open lowlands at Cades Cove and at Oconaluftee, eastern bluebirds feast on insects, spiders, berries, and fruits and build nests in the holes of old fence posts. With their spiked, blue-gray crests, belted kingfishers patrol the streams. They hover over the water and dive in vertically to catch small fish, called stone-rollers, plus tadpoles and crayfish. From hummingbirds and wrens flitting about the understory to ravens and hawks soaring on high, the bird-watching possibilities go on and on like the hills and valleys.

Male American gold-finches exchange their drab winter plumage for bright golden hues in the breeding season.

Creatures Small and Unexpected

Even harder to spot than birds in thick vegetation are the great numbers and various species of toads, frogs, snakes, salamanders, mice, shrews, moles, and voles. Instead of disturbing habitats to find these small creatures, leave these delightful discoveries to chance while enjoying the blossoms and aromas of all the flowering shrubs and wildflowers. A quick movement or a flash of color might catch your eye.

The pine woodlands of the Smokies are the northernmost limit of the green anole. In less than 30 minutes, this lizard can change its coloration from yellowish green to gray to dark brown, depending on the temperature and lighting and the anole's emotional condition. The male anole also inflates a pink throat flap when another male approaches its territory. Not to be outdone in colorful displays, male fence lizards protect their turf by bobbing up and down and flashing their bright-blue undersides and chins.

Of the nearly two dozen species of snakes in the park, only two are poisonous, the timber rattlesnake and the copperhead, and

Dense stands of conifers and deciduous trees (left) provide habitat for 70 species of mammals and 23 species of snakes.

Wolf spiders (right) catch prey by darting out from hiding places rather than by building webs.

Opossums (below), are the only marsupials native to North America. The young are carried in the mother's pouch for up to three months after birth.

they are not known to be aggressive unless disturbed. Timber rattlesnakes occur throughout the park in rocky, heavily vegetated, and grassy areas. They may be black or they may be yellowish with black blotches. Most copperheads, with their rust-brown hourglass markings, are found below 2,500 feet on rocky hillsides, in stone walls, and around old settlements. Seen more commonly are black rat snakes and northern water snakes, which are dark brown. Rough green snakes blend in with the trees and shrubs where they stretch out and rest on the branches, so look again if you see a "branch" wiggle.

Return of the Natives

National parks are established primarily to protect special landforms and native species. Some nonnative plants and animals are absolute pests. Such is the case in the Smokies with the European wild hog, which is rooting up grasses and soil with its sharp tusks and devouring rodents, snakes, salamanders, and other animals. Park managers are attempting to control the damage by relocating some of these exotic hogs. Perhaps they will get some predatory help in the future from the cougar, or panther as it is called in the Smokies. Reports of cougar sightings in the park are becoming more prevalent.

Natural-resource managers are working hard to restore the native brook trout to park waters that have been taken over by rainbow and brown trout. A program to reintroduce the endangered red wolf to replace the nonnative coyote proved unsuccessful and was eventually canceled, but managers are studying the feasibility of reintroducing native elk.

Though most wildlife safarians cannot expect to see the elusive cougar, they can take pleasure in just knowing that these natural predators – along with other native species from black bears to salamanders – thrive in the Great Smoky Mountains, a land of hidden treasures.

TRAVEL TIPS

DETAILS

When to Go

Summer is warm, humid, and frequently wet, with temperatures ranging from the high 80s to the low 50s. Late spring and early fall yield mild days and cool nights, with daytime temperatures in the low 70s and nighttime temperatures in the 40s. Winter is moderately cold and occasionally snowy. Expect changeable weather, especially in the spring. Temperatures in the higher elevations are 12° to 15°F cooler.

How to Get There

The nearest major airport, McGhee-Tyson, is in Knoxville, Tennessee. Asheville Airport is about 60 miles east of the park, in Asheville, North Carolina.

Getting Around

Car rental is available at both airports. The city of Gatlinburg operates a daily trolley shuttle (June to October) to popular park sites. For trolley information, call the Gatlinburg Trolley Department at 423-436-3897. Otherwise, travel within the park is by car or bicycle or on foot.

Backcountry Travel

A free permit is required for overnight travel in the backcountry and may be obtained at campgrounds, visitor centers, and ranger stations. For information, call 423-436-1231.

Handicapped Access

Select trails and campsites are accessible; some assistance may be required. Facilities at Cades Cove, Mountain Farm Museum, Mingus Mill, Cable Mill, and all visitor centers are wheelchair accessible.

INFORMATION

Great Smoky Mountains National Park

107 Park Headquarters Road; Gatlinburg, TN 37738; tel: 423-436-1200.

Bryson City Chamber of Commerce

P.O. Box 509; Bryson City, NC 28713; tel: 800-867-9246 or 828-488-3681.

Gatlinburg Department of Tourism

234 Airport Road; Gatlinburg, TN 37738; tel: 423-430-4148.

CAMPING

The park has more than 1,000 developed campsites at 10 campgrounds. Seven seasonal campgrounds are on a first-come, first-served basis; three are on a reservation system from May 15 to October 31. Reservations can be made three months in advance; call 800-365-2267.

LODGING

PRICE GUIDE – double occupancy

$ = up to $49 $$ = $50-$99

$$$ = $100-$149 $$$$ = $150+

Apple Wood Manor Inn

62 Cumberland Circle; Asheville, NC 28801-1718; tel: 800-442-2197 or 704-254-2244.

This turn-of-the-century, colonial-style house is situated in the historic Montford District of Asheville on 1½ acres of flower gardens, giant oaks, pines, and maples. Spacious guest rooms have antiques, fireplaces, and private baths. Bicycles, badminton, and croquet are available. $$-$$$

Folkestone Inn

101 Folkestone Road; Bryson City, NC 28713; tel: 828-488-2730.

Located a quarter of a mile from

the park, this 1920s mountain farmhouse offers 10 guest rooms, each with private bath, some with a private balcony and mountain vista. The house is furnished with antiques and a collection of mountain memorabilia. Rooms are decorated in different outdoor motifs, such as canoeing, birding, and horseback riding. A large library is stocked with books and maps containing information about area excursions. $$

Fryemont Inn

P.O. Box 459; Bryson City, NC 28713; tel: 800-845-4879 or 828-488-2159.

The Fryemont Inn has lodged guests since 1923. Built by a timber magnate, the inn's exterior is covered with poplar bark, while 37 cozy guest rooms are paneled with chestnut. Rooms have private baths, some with antique pedestal tubs and wrought-iron beds. A mammoth stone fireplace crowns the lobby. A cabin contains two bedrooms with queen-sized beds, kitchen, living room, and baths. Seven cottage suites are located in three buildings and have king-sized beds, fireplaces, kitchenettes, and separate living rooms. Amenities include a restaurant, bar, lounge, and swimming pool. The suites and cabin are open year-round; the main lodge is open from mid-April through October. $$-$$$

Hippensteal Inn

P.O. Box 707; Gatlinburg, TN 37738; tel: 800-527-8110 or 423-436-5761.

This stone-and-wood inn was built in 1990 and is set on 23 scenic acres. Rooms are eclectic, with queen-sized beds, gas fireplaces, private baths, two-person whirlpools, and magnificent views of the Smokies. $$$

LeConte Lodge

250 Apple Valley Road; Sevierville, TN 37862; tel: 423-429-5704.

Five hiking trails lead to this cluster of rustic cabins, perched near the summit of the park's 6,593-foot Mount LeConte. The cabins, built in 1926, are supplied with kerosene lamps, heaters, sheets, and wool blankets. Guests stay in private cabins or private rooms in cabins with shared living rooms. Meals are included in the price and served in the lodge. Open from late March to mid-November. $$

TOURS & OUTFITTERS

Cades Cove Stables

4035 East Lamar Alexander Parkway; Walland, TN 37886; tel: 423-448-6286.

Guided one-hour horseback rides through the foothills of the Smokies. Excursions provide opportunities for observing deer, wild turkeys, black bears, and many small mammals.

English Mountain Llama Treks

738 English Mountain Road; Newport, TN 37821; tel: 800-653-9984 or 423-623-5274.

Llamas carry your gear on day or overnight pack trips. Hearty meals and camping equipment are provided. Excursions can be tailored to your level of experience and fitness.

Great Smoky Mountains Institute at Tremont

9275 Tremont Road; Townsend, TN 37882; tel: 423-448-6709.

The institute conducts workshops for people of all ages. Programs include hiking, slide shows on flora and fauna, living history, and wildlife demonstrations.

Smoky Mountain Adventures

11460 Highway 19; Bryson City, NC 28713; tel: 888-259-5106 or 704-488-2020.

Outdoor programs include day hikes, pack trips, boat tours, and fly-fishing excursions in the park; all are fully outfitted. Destinations range from remote backcountry sites to easily reached spots on Fontana Lake. Pontoon boats are also available.

Excursions

Cherokee National Forest

P.O. Box 2010; Cleveland, TN 37320; tel: 423-476-9700.

Named after the Cherokee Indians, the original stewards of this 635,000-acre area, the forest stretches along Tennessee's eastern border, abutting national forests in Virginia, North Carolina, and Georgia. The state's only national forest contains nine major rivers, a 120-foot waterfall (Bald River Falls), and the nation's first Forest Service Scenic Byway. Many peaks exceed 5,000 feet. Fifty types of mammals inhabit the forest, including wild boars and black bears, and about 120 species of birds.

Pisgah National Forest

1001 Pisgah Highway; Pisgah Forest, NC 28768; tel: 828-877-3350.

Observant visitors may see black bears, wild turkeys, saw-whet owls, magnolia warblers, peregrine falcons, and golden eagles. Whiteside Mountain, a landmark along the eastern continental divide, rises 2,100 feet from the valley floor to a summit of 4,930 feet. Roan Mountain, at nearly 7,000 feet, straddles the North Carolina-Tennessee border and is known for spectacular natural gardens of Catawba rhododendrons. Looking Glass and Moore Cove Falls are both beautiful spots for hikers to cool off.

Shenandoah National Park

P.O. Box 348; Luray, VA 22835; tel: 703-999-2266.

Prior to the park's establishment in 1935, the region suffered from more than a century of widespread farming. The forest diminished, soil thinned, animals disappeared. If replenishing the natural habitat was difficult, acquiring the land was a ponderous task, involving the purchase of more than 3,500 privately owned tracts. Today, however, the forest blankets nearly 200,000 acres. About 100 species of trees shelter a great abundance of bird life. Large mammals include white-tailed deer and black bears.

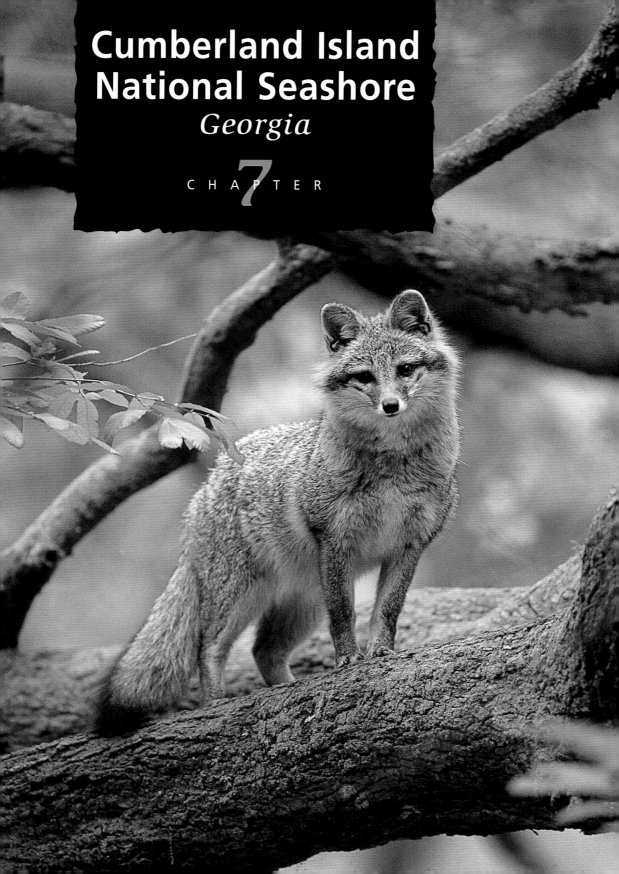

Cumberland Island National Seashore

Georgia

Below a 20-foot bluff overlooking what seems like a continent of salt marsh, a female belted kingfisher is poised to dive into a tidal creek. **Christmas Creek**, winding through the northern end of Georgia's **Cumberland Island**, floods with water so alive that it wiggles and thrashes and mutters as it fills the muddy banks, snapping and popping with shrimp and baby mullet. The kingfisher launches from its perch, flashing as it enters the chartreuse water, and is gone no longer than it takes to read this sentence. Twisting and slinging water, it returns to its perch, shakes the fish in its large bill, and, with a tilt of its black-capped head, jerkily swallows the catch down its gullet.   ◆   A large flock of wading birds knead a mudflat beyond the kingfisher, looking like white water lotus blooming on a black pond. Seen through binoculars, the flowers become white ibis, great egrets, and snowy egrets. In and up the creek and into the marsh the water comes, until it

On this subtropical barrier island, creatures of the forest and shore live by tidal and seasonal rhythms.

seeps between the stalks of spartina. The kingfisher cocks its body and dives again, emerging with a fat silver fish about three inches long. Its dive is propelled by gravity, by the force of the moon, by the same hand that pulls this tidal creek home.   ◆   Saltwater marshes make up nearly a third of Cumberland, the southernmost and largest barrier island on the Georgia coast. These and the forests, shoreline, and surrounding waters sustain considerable wildlife because the 17½-mile-long island is undeveloped, except for a handful of private residences. The entire island was protected in 1972 as **Cumberland Island National Seashore**, and almost 9,000 acres of the northern end, traversed by a network of trails, are designated a wilderness area.

Gray foxes are nimble canids that scamper up trees to survey their surroundings, look for prey, or flee danger.

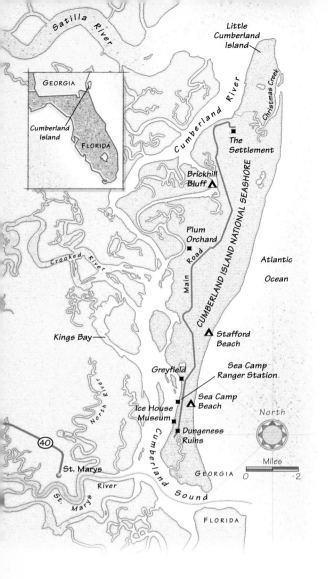

Along this stretch of coast, tidal fluctuations are dramatic, regularly ebbing six to eight feet. At low tide, wading birds move in to feed in the mudflats. Opossums, raccoons, skunks, and other animals sally from the uplands to snare the shellfish and fiddler crabs that scurry in hordes, eating decaying vegetation. The wind is stiff, dragging from the salt marsh a cologne of salty, rich decay.

Only 300 visitors a day are allowed on this island sanctuary, and the only way to reach it is by ferry from the mainland town of **St. Marys**, a small port on the north side of the St. Marys River. The ferry crosses **Cumberland Sound** and docks at Park

Service facilities on the southern end of the island. Summer visitors should expect 100 percent humidity and temperatures in the high 90s. Ticks, mosquitoes, and no-see-ums are alarmingly prolific, so be prepared. In winter, both heat and insects abate. Three venomous snakes – eastern diamondback and timber rattlers, and cottonmouth moccasins – inhabit Cumberland, but they shy away from visitors. Alligators, not likely to cast a second glance at hikers, dally in large, secluded freshwater pools, especially in the north end.

From the ferry dock, it is a short walk inland through maritime forest to the ruins of **Dungeness**, a four-story tabby structure erected by Gen. Nathaniel Greene's widow in the late 18th century. Barn swallows idle around the vine-clad walls that still stand. Feral horses often graze, and bands of wild turkeys forage on the grounds.

The surrounding forest, the upland habitat of this barrier island, is dominated by live oaks, gnomish and ceaselessly amazing in their beauty. Some of their branches form Byzantine arches hung with Spanish moss. Others, as convoluted as a drunkard's path, occasionally rub against each other, creaking through the lonely woods like a great ship meeting waves. Their dark limbs are coated with resurrection fern, so named because it withers and browns in drought, only to be reborn an invigorating green when the rains return. Fluted and needle-tipped fans of saw palmettos rustle in the hammock. Southern red cedar, sparkleberry, and cabbage palm fill the understory. Painted buntings, summer tanagers, cardinals, and pileated woodpeckers flash through the dimness. Carolina wrens sing *teakettle teakettle teakettle,* and yellow-throated warblers call from treetops.

At the island's northern tip, in what's known as the **Settlement**, is the **Cumberland Museum**, where specimens of island fauna are collected and catalogued for use in scientific research. Amphiumas, washed out of freshets and ponds and found dead, are preserved, as are lizards and salamanders. Here is the leg of a wood stork; there an unborn sperm whale from a beached female.

Window on the Atlantic

On the Atlantic side of Cumberland, the ocean flails against the wide, white-sand beach. On summer nights, sea turtles lumber ashore to lay eggs, and the hatchlings emerge from their nests and drift out to sea about eight weeks later. Although several species have been identified on Cumberland, loggerhead turtles, some weighing up to 400 pounds, are most common.

Cumberland is a premier spot on the Atlantic seaboard for observing shorebirds year-round, but particularly before the fall migration. At **Pelican Point**, the extreme south end of the island and a 2½-mile walk from the park dock, rafts of shorebirds – least terns, American oystercatchers, Wilson's plovers – gather in late summer before their journey. Ruddy turnstones and willets poke at the beach with their bills, and royal terns skitter about. Laughing gulls circle, crying hoarsely, before gathering in somber congregations near an American stingray that has recently washed up.

A great blue heron fishes on the frothy strand, near a tricolored heron. A flotilla of brown pelicans flies low over the water. Farther out, the sleek steel bodies of bottle-nosed dolphins sail into the air. Between the beach and the hammock on the windward side are lovely and fragile sand dunes,

two or three rows of them, held in place by sea oats, morning glory, and Spanish bayonet. Many shorebirds nest in hollows atop the sand. (It's important, therefore, to stay off the dunes and use only marked dune crossings.)

The best way to relish Cumberland is to sit still somewhere and watch what passes. A mother raccoon with three juveniles ambles past. Magnolia trees toss and rattle. Rain slips across **Little Cumberland Island**. Five feral pigs – two adults and three young – dig crustaceans from the mud. A red rat snake, flicking its tongue, swirls out of the palmettos.

Cumberland is a world of its own, full of secret life. In this maritime wilderness that humans have not been allowed to sully, the forces of nature continue to push and pull. High and low tides come and go, as do migrations of songbirds and shorebirds. The magnificence and grace of nature are always visible, always unchecked, and never fully known.

Least bitterns (above) can be hard to spot as they climb through dense grass, stalking small fish and aquatic insects.

Feral horses (left) can overgraze the island's vegetation. Cumberland also supports many native species, including alligators, sea turtles, and numerous shorebirds.

TRAVEL TIPS

DETAILS

When to Go

Spring attracts the most visitors, but fall is splendid, with temperatures in the 70s and 80s. Winter on the island is short and mild; temperatures hover around the 50s, with occasional "nor'easters" and brief periods of freezing. Summers are hot, humid, and buggy, with temperatures in the 90s and frequent thunderstorms. There are no shops or restaurants, so bring all necessary food and equipment, and be prepared to pack out your trash.

How to Get There

The island is reached only by boat. Ferries depart daily from St. Marys, Georgia, March through September. The ferry runs five days a week October through February. Island visitation is limited to 300 people per day; reservations are essential and can be made by calling 912-882-4335. Jacksonville International Airport, 30 miles from St. Marys, has car rentals and bus transportation.

Getting Around

Cars are not permitted on the island. Bicycles are allowed on roads but must be brought to the island by private boat. Otherwise, hiking and boating are the only ways to get around.

Backcountry Travel

All visitors are required to purchase day-use and/or backcountry camping permits.

Handicapped Access

The ferry, visitor center, and museum are accessible. A specially designed beach wheelchair is also available.

INFORMATION

Cumberland Island National Seashore

P.O. Box 806; St. Marys, GA 31558; tel: 912-882-4336.

Georgia Tourism

285 Peachtree Center Avenue, Suite 1000; Atlanta, GA 30303; tel: 800-847-4842 or 404-656-3590.

St. Marys Tourism Council

P.O. Box 1291; St. Marys, GA 31558; tel: 800-868-8687.

CAMPING

All camping is limited to seven days and requires both a camping permit and reservation. The park has four backcountry campsites and a 60-person developed campground at Sea Camp Beach, with restrooms, cold showers, and drinking water. To make a reservation, call 912-882-4335.

LODGING

PRICE GUIDE – double occupancy

$ = up to $49 $$ = $50-$99

$$$ = $100-$149 $$$$ = $150+

Cumberland Kings Bay Lodges

603 Sand Bar Drive; St. Marys, GA 31558; tel: 800-831-6664 or 912-882-8900.

A new two-story structure about three miles from the ferry launch. The inn features 116 large and cheery guest rooms with private baths and kitchenettes, as well as a pool, playground, laundry facility, small grocery store, and sports equipment. $

Greyfield Inn

Cumberland Island; P.O. Box 900; Fernandina Beach, FL 32035-0900; tel: 904-261-6408.

The island's only lodging was built in 1901 for Lucy and Thomas Carnegie's daughter, Margaret Ricketson. It was turned into an inn by her daughter, Lucy R. Ferguson, in the 1960s. The four-story house sits on 1,300 private acres and retains its original furniture. Of 11 guest rooms and suites, seven have shared baths. Both of the inn's cottages have two guest rooms with private baths. Cost includes gourmet breakfast, lunch, and dinner. Amenities include a private ferry, guided tours by staff naturalists, bicycle rentals, gift shop, and bar. $$$$

Guest House Inn and Suites

2710 Osborne Road; St. Marys, GA 31588; tel: 800-768-6250 or 912-882-6250.

This newly remodeled brick hotel has 104 guest rooms, each with a king-sized or two double beds, a refrigerator, and microwave. Amenities include a restaurant, lounge, pool, laundry, and playground. $$

Hostel in the Forest

3901 U.S. Highway 82; Brunswick, GA 31521; tel: 912-264-9738 or 912-265-0220 or 912-638-2623.

This unique hostel is set on 95 wooded acres a short drive from St. Marys. Tree houses and geodesic domes offer a combined 40 beds. Hostelers share a communal evening meal, compost toilets, outdoor showers, and occasional chores (including blueberry picking and egg gathering). On the premises are an outdoor hot tub, a common room with library, organic gardens, a glass meditation room, and a pond. $

Riverview Hotel

105 Osborne Street; St. Marys, GA 31558; tel: 912-882-3242.

The hotel was built in 1916 and is set near the Cumberland Island ferry. It has an interesting seashell facade and spacious two-story veranda. Eighteen cozy guest rooms have private baths and one or two double beds; some offer a view of the St. Marys River. A restaurant and lounge are on the premises. $$

Spencer House Inn

200 Osborne Street; St. Marys, GA 31558; tel: 888-840-1872 or 912-882-1872.

This pretty bed-and-breakfast is in the heart of the historic district, just a block from the Cumberland Island Ferry. Built in 1872, the pink, three-story Greek Revival house has verandas on the first and second floors. Fourteen guest rooms, each with private bath, are furnished with antiques. $$

TOURS & OUTFITTERS

Cumberland Island Park Service

107 St. Marys Street West; St. Marys, GA 31558; tel: 912-882-4335.

Ferry trips to Plum Orchard Mansion, where rangers lead a tour, are made the first Sunday of each month. Rangers also lead tours at Dungeness. Tours discuss island ecology and history.

Up the Creek Xpeditions

111 Osborne Street; St. Marys, GA 31558; tel: 912-882-0911.

Guided day tours to Cumberland Island in one- or two-person kayaks; instruction provided.

MUSEUMS

Museum of Cumberland Island National Seashore

129 Osborne Street; St. Marys, GA 31558; tel: 912-882-4336.

The museum features a collection of historic artifacts, ranging from early Native American settlements to the Gilded Age of post-Civil War America.

Excursions

Crystal River National Wildlife Refuge

1502 Southeast Kings Bay Drive; Crystal River, FL 34429; tel: 352-563-2088.

A mere 46 acres, this refuge holds monumental importance for approximately 200 manatees. The warm (72°F) waters of the Crystal River are a winter sanctuary for these gentle creatures, which require water temperatures no colder than 68°F. Though the refuge was founded in 1983 to protect manatees, it is home to about 250 species of birds and a number of reptile, amphibian, and mammal species, including black bears and bobcats, two of Florida's rarest creatures.

Okefenokee National Wildlife Refuge

Route 2, Box 338; Folkston, GA 31537; tel: 912-496-3331.

This primitive 395,000-acre refuge is home to more than 10,000 alligators. Habitats range from cypress and tupelo blackwater swamp to freshwater marsh and pine uplands. A great diversity of creatures thrive here, including one of the largest black bear populations in the South. River otters are readily seen by canoeists. Of the shelter's many aquatic birds, the tricolored heron, sandhill crane, and white ibis stand out. Woodpeckers, songbirds, and raptors abound.

St. Marks National Wildlife Refuge

P.O. Box 68; St. Marks, FL 32355; tel: 904-925-6121.

The jaguarundi, a dark, slender wildcat seldom found north of Mexico, ranges this 100,000-acre refuge. St. Marks includes 35,000 acres of water and 17,000 acres of designated wilderness area. The refuge is inhabited by a rich variety of wildlife – from wild turkeys and bobcats to sea turtles and manatees. According to the Audubon Society, this is one of the nation's most "birdy" places. Scores of waterfowl spend the winter here, turkeys strut through the woods, and bald eagles and ospreys are often seen fishing.

Everglades
National Park
Florida

Dozens of wood storks crowd the mangrove swamp across **Paurotis Pond**, like Christmas lights among clouds of green leaves. One occasionally rises, pulling into the air with five feet of wings. It flies out across the mangroves, then returns to perch again, careful not to crowd the other storks or the great egrets and white ibis dotted among them. In the pond itself, a long, dark-brown snout, followed by a pair of eyes, eases through water as warm and dark as coffee. The eight-foot-long alligator slides out of the pond through a break in the mangrove roots and climbs into the sunshine. ◆ Wading birds feed at the pond's muddy edge: glossy ibis with immense decurved bills, tricolored herons outlined in white, and snowy egrets wearing golden slippers. In the mellow, slanted sun, roseate spoonbills, their epaulets like red velvet, snatch small fish with yellow-green bills. Departing in small groups, the wood storks awkwardly take to wing, flapping out over the swamp. They head north to sawgrass prairies where they will

Alligators wallow, frigatebirds dive, and manatees roll in the wildlife-rich and ecologically fragile "river of grass."

feed on mosquito fish and sunfin mollies in the shallow pools left by winter's dry season. ◆ Paurotis Pond, named for a variety of palm, lies deep in the **Everglades**. Because of the influences of wind and sea, this vast subtropical wilderness at the southern tip of Florida functions like the tropics. Water is the element that shaped the landscape, and on it the land depends, like a long-married couple that cannot survive without each other.

Alligators, a common sight in the Everglades, have broader snouts than crocodiles, which are found only at the southern tip of Florida.

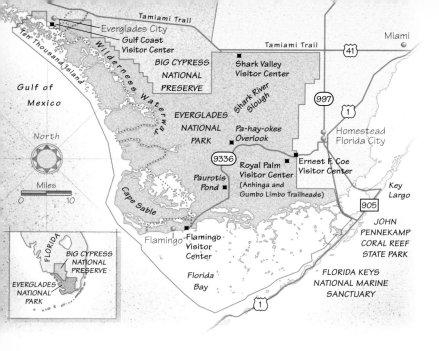

Map labels:
- Tamiami Trail
- Everglades City
- Gulf Coast Visitor Center
- BIG CYPRESS NATIONAL PRESERVE
- Shark Valley Visitor Center
- Tamiami Trail
- Miami
- 41
- Ten Thousand Island
- Wilderness Waterway
- Shark River Slough
- 997
- 1
- Gulf of Mexico
- EVERGLADES NATIONAL PARK
- Pa-hay-okee Overlook
- Homestead Florida City
- North
- 9336
- Royal Palm Visitor Center (Anhinga and Gumbo Limbo Trailheads)
- Ernest F. Coe Visitor Center
- Paurotis Pond
- Key Largo
- 905
- Miles 0 10
- Cape Sable
- JOHN PENNEKAMP CORAL REEF STATE PARK
- Flamingo
- Flamingo Visitor Center
- FLORIDA KEYS NATIONAL MARINE SANCTUARY
- Florida Bay
- FLORIDA
- BIG CYPRESS NATIONAL PRESERVE
- EVERGLADES NATIONAL PARK
- 1

Water Course

The Everglades, now confined to a 1½-million-acre national park established in 1947, is just one-fifth of its original size. Geographically, the Everglades ecosystem begins in the environs of **Lake Okeechobee**, the second largest freshwater lake in the United States. At one time, as the lake overflowed its southern rim, the flood water, swollen by rain, embarked on a slow journey across the slightly tilted landscape of southern Florida. Today, this process happens in a less spontaneous, less natural way. A network of canals and impoundments metes out water for the region: some to Miami, some to sugarcane farmers, some to the Everglades. What water there is inches through sawgrass prairie and marsh, through fresh-water sloughs and cypress bays, filling alligator wallows and marl pools. It floods mangrove estuaries – some of the most productive ecosystems in the world – and coastal prairies, with their coverlets of yellow-green saltwort, finally joining the life-rich **Florida Bay**, land of 10,000 keys.

The name "Everglades" stems from unending prairies of sawgrass, a tawny sedge that can grow to 12 feet, whose quarter-inch blades are lined with sharp serrations. The wet, grass-covered plain seems to stretch forever, broken only by tree islands. Even the Seminole Indians who lived here (their reservation is northwest of the park) called these southlands Pa-hay-okee, or "grassy water." It was the author and activist Marjory Stoneman Douglas who called it a river – a river of grass – only inches deep and 50 miles wide, following a haphazard course 100 miles long.

Edward O. Wilson, in his book *Biophilia*, theorizes that we humans are attracted to savannas because that is the landscape – the African plain – of our origin. Perhaps the lure of the Everglades is the unmistakable feeling that life begins on this tableland, springing from the fantastically pitted limestone floor, and from the peat-thick middens of the hammocks, where earth and water meet. What cannot be learned here about genesis, about the human place in the landscape and the aim of life, cannot be learned at all.

Anhingas (left) spear fish with their sharp bills, then eat them head first.

Peninsula cooters (opposite, above) and other turtles sun themselves on logs and stones near the park's boardwalk trails.

Corn snakes (opposite) are nocturnally active constrictors. Their name comes from the similarity of their colors to those of Indian corn.

Snake Birds and Living Jewels

The dry season (November through April) is the time to visit the Everglades, because lower water tables concentrate wildlife in and around ponds, creating spectacular displays. In these concentrations, it is possible to observe phenomenal acts of natural history – a great blue heron pounding a large fish or a crow struggling with a yard-long snake. There are other reasons to visit in winter. It is so humid and hot in summer that you want to shed your own skin, and the biting insects, especially mosquitoes, are buzzing and boiling by the thousands. The bugs abate in winter, but never disappear. Summer is also rainy season, when thunderstorms are likely. Hurricane season officially begins in June, though most arrive from late August to October.

One road, the 38-mile Main Park Road, cuts through Everglades National Park from **Florida City**, south of **Homestead**, to **Flamingo**, following the water's path. Strung along the road are hiking trails, observation towers, boardwalks, and other viewing sites. Not far from park headquarters, the **Anhinga Trail** winds alongside **Taylor Slough** and its cattail marsh, a major drainage of the Glades and one that remains saturated even through the dry season. Along Anhinga Trail, the namesake bird, known as "snake bird" for its habit of swimming with only its long neck and head exposed, dries its wings almost within reach. Among drooping willows, white ibis wade, curtsying to purple gallinules balanced on water lily pads. The gallinules, with bills that are part red and part yellow, are as stunning as the name suggests. Lilies bloom all across the slough, and fluffy white balls of wide-ranging buttonbush call to the viceroys, which open on the flowers like little orange books. Alligators bask in the sun, and red-winged blackbirds chirr from the cattails, flanked

by ghostly white spider lilies and purple spires of pickerelweed.

Nearby **Gumbo Limbo Trail** leads through a tropical hardwood hammock dense with pigeon plum, Jamaica dogwood, mastic, strangler fig, wild coffee, ironwood, and stopper. White-crowned pigeons, unique to south Florida, flock to these junglelike tree islands in summer to eat fruit. Raccoons, white-tailed deer, cotton mice, and green treefrogs forage here. The trees crawl with gorgeous multicolored tree snails, two to three inches long, known as the "living jewels" of tropical hardwood hammocks. The snails, in whorls of emerald green, brown, orange, yellow, or pink, scrape algae and

fungus from smooth-barked trees. They estivate during the dry season and are most active from May through September.

The Everglades is subtle, and the only way to its heart is by water, which means getting wet. It also means being quiet and deliberate, maybe contemplative, slowing down enough to see the otter at play. Stop and slog through the sawgrass toward one of the teardrop-shaped tree islands. The marl in the calf-deep water swirls and clouds like wet wood ashes on the ancient sea bottom. Sawgrass skippers flutter by. A red-shouldered hawk banks in the sky, and boat-tailed grackles complain. Near the tree island, a yellowthroat calls *witchity, witchity* between the oinks of a pig frog. Out across the prairie, a pair of sandhill cranes rummage through the water life. In the sky, fat clouds pile up like blown cotton.

Wade into the dwarf cypress swamps, hung with Spanish moss and adorned with bromeliads. Though diminutive, the trees are ancient. This is the place to be at sunset and moonrise. In winter, the leafless cypress appear dead, but their green, feathery foliage emerges again in spring.

Kites and Crocs

The park road continues through pine rocklands, one of the most critically endangered ecosystems on the continent. These mark the highest elevations, sometimes 10 feet above sea level, but in a land this flat, mere inches seem like mountains, so dramatic is the change in vegetation. Here, slash pines, along with an astounding number of other endemic flora, including an understory of saw palmetto, root in the cracks of lime-

Marjory Stoneman Douglas

It was in 1947, the year President Harry S. Truman dedicated Everglades National Park, that Marjory Stoneman Douglas published *Everglades: River of Grass*. Telling in lyrical prose the story of the area's ecology, wildlife, and human history, the book did more to protect the misunderstood Everglades than any other single action in history.

Douglas was born to a long line of strong-willed, outspoken individuals – Revolutionary War soldiers, abolitionists – who had worked for social change. In 1915, after a failed marriage, she joined her father, Judge Frank Stoneman, in Miami. Founder and editor of the *Miami Herald,* he recognized the value of the Everglades, which most people dismissed as a worthless, snake-infested swamp, and he criticized the state's plans to drain it. Douglas worked for the paper before striking out on her own as a writer.

In her book's last chapter, "The Eleventh Hour," Douglas calls for beauty to win over greed in determining the future of the Everglades. It wasn't until she was turning 80 that she became an environmental activist. She created the Friends of the Everglades and led it in defending the beleaguered, development-pressed landscape. Hers was a powerful, uncompromising voice for protection and restoration. Marjory Stoneman Douglas died in 1998 at the remarkable age of 107.

A wilderness area within the national park is named after Marjory Stoneman Douglas (left) in recognition of her conservation efforts.

Great blue herons (opposite) build large nests on the ground or high in trees for protection from predators.

Orange sulphurs (below), also known as alfalfa butterflies, begin life as a pink-striped green caterpillar that forms a green chrysalis.

---

stone outcroppings. In summer, swallow-tailed kites soar across the pine, banking and circling. Their sleek black-and-white body and deeply forked tail seem the most graceful things on earth. The birds perch regally on pine limbs, their tails like closed Japanese fans.

At peninsula's end is **Flamingo**, which a century ago was a barely accessible town of less than 100 people. From here, tourist boats leave regularly for day excursions into **Florida Bay**. If the tide is out, all kinds of sea and wading birds will be probing the mudflats. Look also for herds of fiddler crabs. Elusive crocodiles, more scarce than alligators, curl like commas on the exposed banks. They are lighter in color than their cousins, an olive gray, with narrow snouts and lower teeth, especially the fourth, clearly visible. Thanks to protection efforts, numbers of crocodiles in the warm waters of southern Florida are increasing, from a low of 300 in the 1970s.

Among platinum ripples in the bay, bottlenose dolphins cavort, flanking and following boats. A peregrine falcon perches in a mangrove. Magnificent frigatebirds, glossy to brownish black, skim the bay, snatching up fish they encounter at the surface; they don't swim because takeoff from the water is hampered by small feet and short legs. In the air, they sometimes harass other birds into regurgitating food, which they promptly pirate. During courtship, males enact a remarkable ceremony: rapid vibrations of wings and inflation of an orange throat pouch that flames bright red.

From Flamingo, a 99-mile canoe trail called the **Wilderness Waterway** runs northwest to **Everglades City**. The waterway is open to serious backcountry canoeists as well as to visitors on motorized day excursions.

Manatees (left), the only completely vegetarian marine mammal, live in the warm, shallow waters of the Ten Thousand Islands.

Florida panthers (right), a sub-species of mountain lion, are rarely seen. Only a few dozen of the endangered animals are believed to inhabit south Florida.

It is a green labyrinth of impenetrable-looking tunnels through one of the largest mangrove forests in the United States.

Mangroves colonize tidal flats. The cigar-like seeds of the pioneer species, red mangrove, float in salt water until the roots touch bottom and take hold. Called "walking trees," they are noted by their prop roots, which give them the appearance of standing on tall legs. Mangroves serve as nurseries for more than 80 percent of south Florida's fish and invertebrates, including spiny lobster, pink shrimp, mullet, tarpon, snook, mangrove snapper. Above water, mangrove squirrels scamper in the tangled limbs; anhingas and cormorants frequent them, as do mangrove cuckoos, which sit quietly near the heart of the tree. The air is saturated with the smell of decomposition, life becoming life.

Sea Cows

At **Everglades City**, portal to seekers of true wilderness, the tannic-stained bay opens into **Ten Thousand Islands**, the most remote corner of Florida and a world all its own. From the air, the islands are like green marbles set into a deep-blue basin. Around their edges the current swirls, combing out blond tails of sand. The best way to see this wilderness is by kayak, striking out from the marina at Everglades City just at high tide, compass and nautical chart within reach, crossing the bay into the maze of uninhabited keys.

Between islands, bottlenose dolphins fill the silence with their leaps and misty exhalations. Flotillas of brown pelicans fly low. Osprey build ragged nests in the mangroves, as do bald eagles. These commonly confused birds of prey hunt by torpedoing their bodies into the bay, coming up in a shower of slung water, fish in talons.

Occasionally a manatee, or sea cow, rises to the surface to blow, revealing an immensity startling in this estuary, where large animals are mostly absent. Manatees are shy, gentle, vulnerable vegetarians that feed in the sea-grass beds. Curious by nature, they often investigate human activities, which sometimes prove harmful. They can't escape boat propellers and are susceptible to fatal infections of untreated wounds.

About 1,500 manatees remain in this country, and the sight of one is a small but not unlikely miracle. To the hopeful, anything that surfaces appears first to be a manatee, but often transforms into a shark or a ray, or the fine flashing fin of a tarpon. But sometimes, in a land that never fails to surprise, sometimes it actually is a manatee, rolling and floating in water that has come so far. The water began as rain and eased its way along the beautiful length of southwest Florida, bringing life with it, absorbing the feel of a particular grace – a primal one that lodges somewhere near your ribs – known as the Everglades. It is worth joining that flow.

TRAVEL TIPS

DETAILS

When to Go

Peak visitation occurs during the park's dry season, mid-December through mid-April, when temperatures average in the 70s and 80s. The rainy season, May through November, is hot, humid, and prone to thunderstorms, with temperatures in the 80s and 90s. Mosquitoes are relentless at this time of year. Insect repellent is essential.

How to Get There

Miami International Airport is about 50 miles from the visitor center near Homestead. Car rentals are available at the airport. Greyhound, 800-231-2222, provides bus service between the airport and Homestead.

Backcountry Travel

The backcountry is accessible by boat or on foot. Backcountry permits are required and may be purchased at the ranger stations and visitor centers no more than 24 hours in advance.

Handicapped Access

Visitor centers, tram tours, several campsites, and most self-guided nature trails are accessible.

INFORMATION

Everglades National Park

40001 State Road 9336; Homestead, FL 33034; tel: 305-242-7700.

Florida Division of Tourism

126 West Van Buren Street; Tallahassee, FL 32399; tel: 904-487-1462.

Homestead/Florida City Chamber of Commerce

43 North Krome Avenue, 2nd Floor; Homestead, FL 33030; tel: 888-352-4891 or 305-247-2332.

CAMPING

The park has three campgrounds – Lone Pine Key, Flamingo, and Chekika – and 362 campsites. Camping is offered on a first-come, first-served basis May to October, and by reservation November to April. To make a reservation, call 800-365-2267.

LODGING

PRICE GUIDE – double occupancy

$ = up to $49 $$ = $50-$99

$$$ = $100-$149 $$$$ = $150+

Best Western Gateway to the Keys

411 South Krome Avenue; Florida City, FL 33034; tel: 305-246-5100 or 888-981-5100.

Each of the motel's 114 guest rooms is spacious and breezy. Standard rooms feature two queen-sized beds or one king-sized bed. Mini-suites and king-bedrooms have a wet bar, refrigerator, microwave, and coffee maker. A pool and spa are on the premises. $$-$$$

Everglades International Hostel

20 Southwest 2nd Avenue; Florida City, FL 33034; tel: 800-372-3874 or 305-248-1122.

This hostel occupies a former Art Deco boardinghouse that escaped the fury of Hurricane Andrew. Just 15 minutes from Everglades National Park and 20 minutes from John Pennekamp Coral Reef State Park, the hostel has two private rooms, a family room, several single-sex and coed dorms, and common sitting rooms. A fully equipped kitchen is stocked with spices, oils, and other cooking supplies; a large tropical garden has a gazebo. Amenities include canoe rentals and free Internet access. $

Flamingo Lodge, Marina and Outpost

1 Flamingo Lodge Highway; Flamingo, FL 33034; tel: 800-600-3813 or 941-695-3101.

Situated in the southern tip of the park, this wilderness resort offers 103 motel rooms and 24 duplex cottages. The motel's six wood-frame buildings, built in the 1950s, offer comfortable rooms with private bath. Each cottage has a full kitchen, two double beds, sitting room, and private bath. Canoes, kayaks, and bicycles may be rented, and a pool is available. $$-$$$

Ivey House

107 Camellia Street, P.O. Box 5038; Everglades City, FL 34139; tel: 941-695-3299 (November to April) or 860-739-0791.

A former boardinghouse for men working on the Tamiami Trail in the 1920s, this bed-and-breakfast has 11 simple guest rooms, most with shared bath, all furnished with turn-of-the-century antiques. Free bicycles and a coin laundry are also available. Closed May through October. $-$$$.

Katy's Place

31850 Southwest 195th Avenue; Homestead, FL 33030; tel: 305-247-0201.

This contemporary two-story bed-and-breakfast is surrounded by an acre of lush tropical plants. The inn has three large guest rooms, two with queen-sized beds and shared bath, one with king-sized bed and private bath. Rooms feature Victorian decor and antiques. Extras include a pool, Jacuzzi, and pond with tropical fish. $$-$$$

Room at the Inn

15830 Southwest 240th Street; Homestead, FL 33031; tel: 305-246-0492.

This country ranch house, with a facade of coral rock, is set on two quiet, rustic acres. Four guest rooms are handsomely appointed with period antiques. Three rooms have queen-sized beds and private baths, one has a double bed. A cedar cathedral ceiling towers above the sitting room, which has a large stone fireplace. Amenities include a pool, heated spa, sun deck, and wet bar. $$-$$$

TOURS & OUTFITTERS

Caribbean Watersports Enviro-Tours

P.O. Box 781, MM 97; Key Largo, FL 33037; tel: 800-223-6728 or 305-451-3595.

Flexible itineraries allow for naturalist-led rafting excursions into isolated regions of both the Florida Keys and the Everglades. Frequently seen wildlife include manatees, dolphins, bald eagles, ospreys, and a great variety of wading birds.

Everglades National Park

40001 State Road 9336; Homestead, FL 33034; tel: 305-242-7700.

Park rangers guide nature hikes, canoe excursions, and tram tours.

Flamingo Lodge Marina and Outpost

1 Flamingo Lodge Highway; Flamingo, FL 33034; tel: 800-600-3813 or 941-695-3101.

The lodge offers a variety of services, including narrated boat tours into Florida Bay or the backcountry, rentals of canoes, skiffs, and bicycles, and fishing charters.

North American Canoe Tours

107 Camellia Street; Everglades City, FL 34139; tel: 941-695-3299 or 941-695-4666.

These fully outfitted canoe expeditions range from two to six nights; some include stays at a bed-and-breakfast. Canoes, tents, fresh-cooked meals, and backcountry camping permits are provided.

Excursions

Big Cypress National Preserve

HCR 61, Box 110; Ochopee, FL 33943; tel: 941-695-2000.

Fourteen endangered species live in this preserve adjacent to Everglades National Park. Together the two parks are the continent's largest breeding ground of tropical wading birds. Alligators, bald eagles, ospreys, and manatees are commonly spotted. Wading birds may be seen in both rookeries and feeding areas.

J.N. "Ding" Darling National Wildlife Refuge

1 Wildlife Drive; Sanibel, FL 33957; tel: 813-472-1100.

This 4,975-acre refuge on Sanibel Island, near Fort Myers, is home to flora and fauna of both subtropical and temperate climate zones. More than 290 bird species are attracted to brackish and saltwater habitats, including many types of wading birds, raptors, and migratory songbirds. An excellent canoe trail leads paddlers through the refuge.

National Key Deer Refuge

P.O. Box 510; Big Pine Key, FL 33043; tel: 305-872-2239.

The diminutive Florida Key deer – a subspecies of the white-tailed deer – was driven nearly to extinction before the establishment of this refuge on several keys in 1957. The deer stand about two feet high and are easily spotted. Other endangered creatures in the refuge include Hawksbill and loggerhead sea turtles, Stock Island tree snails, and yellow-and-black-striped Schaus' swallowtail butterflies.

John Pennekamp
Coral Reef
Florida

CHAPTER **9**

On a balmy Florida afternoon, a party of scuba divers sinks beneath the clear tropical seas some six miles off Key Largo, like Alice slipping through the looking glass. Down they go in a slow-motion free fall, aiming for the living city of coral below. The sweeping ecological community of a coral reef named **Molasses** at first appears as a kaleidoscope, a shifting mosaic of color and form. As their eyes gradually become accustomed to the muted blue light, the underside of the looking glass begins to take shape.   ◆   The terrestrial landscape seems mirrored in a vague, abstract way by the corals. Thick elkhorn branches up toward the surface in imitation of ancient tree trunks, while staghorn hunkers down like thorny bushes in the understory. In deeper water, haystack-sized mounds of star coral lie scattered like giant boulders. Soft corals, waving in the light current like fields of grain, are configured as plumes and whips and finely latticed fans.   ◆   In this liquid underworld, Technicolor tropical fish undulate in waves or poke cryptically about in the reef's fretwork. Many are named as living sight gags for the things they most resemble back in the air-breathing world: butterfly fish and parrotfish, cardinalfish and houndfish, seahorse and burrfish. Others seem to embody fragments of freshwater critters like parts of half-remembered dreams: the Nassau grouper with its low-slung jaw hints of a largemouth bass; the pug-nosed cocoa damselfish recalls a palm-sized bluegill.   ◆   Then there are creatures here with no land reference at all: the green moray eel panting with an angry slit of a mouth; the spiny lobster waving its antennae as if telegraphing a signal; the

Vibrant corals and marine critters beckon wildlife watchers to explore a tropical underworld in the Florida Keys.

Scuba diving and snorkeling are the best ways to see the variety of creatures that live in the coral reef community. Algae give corals their vibrant colors.

A Fragile World

Coral reef communities worldwide are at risk and easily damaged by careless visitors. Here are some do's and don'ts:

• Never anchor on a living reef. Instead, tie up to mooring buoys or set the anchor in the sand. Throughout the entire Florida Keys National Marine Sanctuary, damaging coral will bring a stiff fine.

• Never purposely touch any part of the reef. Loss of the protective mucous on corals opens them to infection, just like a cut on your skin.

Corals (above) are fragile, living structures easily damaged by even a diver's touch.

Brain coral (below) has a surface that resembles the human organ. The shapes of other corals, such as star and elkhorn, also look like their namesakes.

Christ of the Deep (opposite), an undersea landmark, stands in the waters of the Florida Keys National Marine Sanctuary, just beyond the park's edge.

• Be properly weighted when diving so you don't sink like a rock atop the corals. Learn good buoyancy compensation so you can "hover" in place rather than rise or fall with flailing flipper kicks, which can stir up sediment that smothers corals.

• Don't disturb any creature that appears immobile. Nurse sharks, usually docile, will bite if provoked; bristle worms will sting, as will scorpion fish.

• Learn more about the natural history of the reef so your visit will be a more rewarding and engaging one. Remember, you are a privileged visitor to a rare and fragile sanctuary – take only pictures, leave only bubbles.

discus-shaped French angelfish glowing almost iridescently, like an old Jim Morrison poster under black light. In the lee of a bright purple sea fan, thousands of nearly transparent young fry swarm, each no bigger than a single letter on this page. On the surface of the coral, orange Christmas tree worms zip in and out of their holes like whirling dervishes.

Through it all, the exhaled bubbles of the divers rise like domes of mercury and trail back to the surface, balloons of air with no casings. It is a reminder that, despite their immersion in this foreign environment, the divers are still surface-breathing mammals and their visit here is

ephemeral, limited to the amount of compressed air they carry in tanks on their backs.

A Protected Sanctuary

This city of coral, the largest living coral reef system in the continental United States, is a place where warm, relatively clear and shallow waters nurture a living "barrier" of corals from four to six miles offshore, paralleling the island chain of the Florida Keys from just north of Key Largo to the Dry Tortugas. This matrix of coral stretches for nearly 220 miles, making it the third largest such system in the world after the reefs of Australia and Belize.

The best of what this natural system is all about is found seaward of 25-mile-long **Key Largo**, under 15 to 60 feet of water. There are two very good reasons for this, one natural, the other man-made:

• This elongated island at the top of the archipelago has buffered the cooler, sediment-laden upland water that would otherwise wash directly onto the reef from the westerly Everglades and Florida Bay. In doing so, it shelters the tropical corals, which need warm, clear water to grow.

• The reefs just east of Key Largo have been "managed" separately as the **John Pennekamp Coral Reef State Park** since 1960 (the Key Largo National Marine Sanctuary expanded that protection into deeper federal waters in 1975, and 15 years later, the entire Keys were put under federal management). As a result, coral collecting and spearfishing were outlawed here long before it was popular to do so. Corals are generally healthier and fish larger and more diverse than elsewhere along the reef tract.

More a perforated series of coral ridges than a solid barrier, individual sections of this reef have been named and charted by local fishermen, sailors, and divers for years. Most dive sites are described for a geographic characteristic (Grecian Rocks), or a shipwreck (the *Benwood*), or a creature (Conch). Historic century-old steel lights mark where the most prominent reef sections rise to within feet of the surface, from **Carysfort** in the north, to the distinctive middle bend in the reef line at **The Elbow**, to **Molasses** in the south. Some sites are punctuated with temple-like mounds of classic "spur and groove" formations, ridges of coral fissured with sand valleys from the east-west surge of the tides.

If this reef sometimes appears artistically crafted, it is the cast of a sculpture that has become the single most popular site for sub-aquatic visitors. Here the *Christ of the Deep* imitates *Il Cristo Degli Abissi* (Christ of the Abysses), which rests on the sea floor off Genoa, Italy. Made of solid bronze, the nine-foot-high Keys statue was set in place by the Underwater Society of America in the 1960s. This *Christ of the Deep* stands upright in 25 feet of water, arms outstretched and head back, acknowledging all who have ever come to the sea.

Irreplaceable Diversity

While northerners flock here in the winter to escape cold weather back home, the best underwater season actually is summer and early fall when the choppier, cooler seas give way to waters that are warmer and, often, pond-flat-perfect for novice divers and snorkelers. There is plenty here for advanced divers, too. Veterans often sustain their interest by learning underwater photography. Or they engage in a vicarious "hunt" for as many types of fish as they can find, keeping records of them as birders do with their feathered quarries back on land.

Indeed, it is the sheer biological diversity of the reef that leaves most underwater visitors in awe. For instance, the notion of

looking up into a tree and seeing, say, 50 or 60 different species of avifauna would send most ornithologists – amateur and otherwise – into a swoon. That's how many different kinds of fish an astute diver is likely to identify on a coral reef. A 1990 survey conducted by the nonprofit Key Largo-based REEF (Reef Environmental Education Foundation) inventoried some 205 different fish just offshore.

If all the living marine creatures are counted – from corals to fish to invertebrates like sea cucumbers, lobster, and crabs – up to 3,000 different life-forms might be found on a single healthy reef. No wonder biologist and Pulitzer Prize-winning author Edward O. Wilson, who studied the islands here in the 1960s, described coral reefs as "the marine equivalent of rainforests." As a result, Wilson said, this singular North American reef should forever be protected. The warm waters of the Gulf Stream help make this so. They flow out of toasty southern latitudes, paralleling the coast of Florida and helping to incubate this tropical reef system with temperatures that average from 75° to 86°F.

What Is Coral?

Once believed to be a plant, the coral itself is a complex little animal that functions a bit like a sea anemone – an animal to which it is related, for both are anchored to the sea floor and grouped in the phylum Cnidaria. Look closely at the reef and you will see that it is made up of millions of these tiny polyps, some no bigger than the nail on your little finger. Each lives inside a rocky cup, a home of calcium carbonate that it secretes from minerals it draws from the sea. The soft tissues of each cup are lined with a symbiotic algae called zooxanthellae. It is this algae that not only nurtures the polyp with oxygen transferred from the energy of the sun, but also imbues the reef with vivid oranges and reds, greens, and browns. Add cup upon rocky cup and you have a coral reef, a living wall that creeps toward the light at the rate of about an inch a year.

By day, the tiny tentaclelike arms of the polyp usually curl up inside the cup in which it lives. But by night, the tentacles extend to feed, like petals opening on a flower, snacking on hapless animal and plant plankton that strays too close. For divers venturing out on this reef after dark, the normally cheerful world of sunlight and color is transformed: Some hyperactive daytime creatures, like the buck-tooth parrotfish, lie motionless and asleep inside a virtual cocoon spun of mucous. More secretive species, like the copper-colored glassy sweepers, boldly venture out from hiding places inside caves and wrecks. The diurnally sluggish nurse shark, an elasmobranch equipped with cat-fishlike barbules, swirls over the sea bottom in search of dinner. From their daytime burrows come the spiny lobsters, even the moray eels. When divers switch off their underwater lights, bioluminescing plankton may flash around them in the dark sea like Christmas lights, drawing on the same blue-green energy as terrestrial fireflies.

The best way to sort out one creature from another in this strange land is with a handy field guide. The *Marine Life Identification* series by Paul Humann is far and away the best. But there are also "behaviors" to watch for that are equally fascinating: the defensiveness of the striped sergeant major damselfish as it guards its patch of newly deposited eggs; the "cleaning stations" where larger grouper and snapper wait in line to have their gums and gills scoured of parasites by sliver-sized gobies; and the telltale habit of the octopus in leaving empty seashells at the mouth of his den, the octo-pus's garden in the sea.

Coral spawning, a rare, once-a-year nocturnal spectacle, can even be witnessed by those divers in the right place at the right time – usually in August or September, when the superheated waters and the lunar cycle trigger the spawn. During such events, polyps simultaneously fire out millions of tiny BB-sized packets of eggs and sperm that briefly fill the the sea with an opalescent blue-pink snowstorm. Off this spawn goes into the night, drifting sometimes miles before settling to grow new polyps, the genesis of a reef-to-be.

But a coral reef is more than a visual feast, more even than a habitat-rich oasis in a vast sea bottom that is little more than a desert of sand. Take away this reef and you would eliminate a natural underwater bastion that helps diffuse the energy of the incoming sea, leaving low coastal shores and islands exposed to erosion from wind and tide. Without the living fence of the reef, the low-lying Keys would be inundated by the time-less oceanic swash, a reality that takes this world of magic and wonder on the other side of the look-ing glass and turns it into ever-lasting utility.

Diversity (opposite, above) is the hall-mark of a healthy reef. More species of wildlife can readily be seen in the park's reefs than on the nearby coast.

Queen angelfish (opposite, below), a brilliantly colored species that is as flat as a disk, cruise the reefs, feeding on sponges.

Proper training is essential for safe diving (right). First-time visitors who are informed and prepared will enjoy diving in the park's waters.

Rules of Safe Diving

● Train with one of the national certifying agencies in scuba diving.

● Always dive with a buddy.

● Never dive in conditions beyond your skill level.

● Never hold your breath underwater or ascend faster than your exhaust bubbles.

● Know how to read and understand decompression tables. Plan your dive by how long you intend to stay at a certain depth.

● Stay in reasonable physical shape. Maximize your "bottom time" underwater by aerobic exercises that help slow your breathing rate, allowing you to become more relaxed and better able to enjoy your dive.

TRAVEL TIPS

DETAILS

When to Go

The peak visiting season runs from December through April, when there is less rain, cooler weather, and fewer mosquitoes. Temperatures at this time range from the high 60s to the low 80s. Summer temperatures routinely soar into the 90s.

How to Get There

Miami International Airport is 60 miles from the park. Car rentals are available.

Getting Around

Diving tours, charters, and boat rentals are available in the town of Key Largo. For information about licensed outfitters and tours, call the Key Largo Chamber of Commerce at 305-451-1414. Coral Reef Park Company, 305-451-1621, offers glass-bottom boat tours from Key Largo.

Handicapped Access

All buildings and trails are accessible.

INFORMATION

John Pennekamp Coral Reef State Park

P.O. Box 487; Key Largo, FL 33037; tel: 305-451-1202.

Key Largo Chamber of Commerce

105950 Overseas Highway; Key Largo, FL 33037; tel: 305-451-1414.

Florida Division of Tourism

126 West Van Buren Street; Tallahassee, FL 32399; tel: 904-487-1462.

CAMPING

The park has one campground, with 47 tent and RV sites, restrooms, and showers. Nearly half of the sites are available by reservation, which may be made up to 11 months in advance. For information, call 305-451-1202.

LODGING

> **PRICE GUIDE** – double occupancy
>
> $ = up to $49 $$ = $50-$99
>
> $$$ = $100-$149 $$$$ = $150+

Bay Harbour Lodge

97702 Overseas Highway; Key Largo, FL 33037; tel: 800-385-0986 or 305-852-5695.

This rustic complex on a small private beach includes a cluster of bungalows and a motel. All rooms have Mexican tiles and a tropical atmosphere; some have full kitchens. The motel has eight rooms, five with queen-sized beds, three with doubles. Two bungalows are poolside, four overlook Florida Bay. Paddle boats, canoes, and kayaks are available. $–$$$$

Jules' Undersea Lodge

51 Shoreland Drive; Key Largo, FL 33037; tel: 305-451-2353.

Guests at this unique inn can't dive into bed until they've swum to the lodge. Once a scientific research facility, this underwater habitation was turned into a lodge in the 1980s. Two snug, modern, and comfortable guest rooms offer aquatic views through 42-inch windows. Guests share a bathroom. Equipment rentals and diving lessons are available. $$$$

Largo Lodge

101740 Overseas Highway; Key Largo, FL 33037-2664; tel: 800-468-4378 or 305-451-0424.

Less than a mile from the park, this lodge's six duplex apartments offer motel-style rooms. The one-bedroom units have two double beds and a kitchen, living room,

private bath, and screened porch. A dock, shaded by palm trees and lined with lounge chairs, extends over the gulf from the private beach. $$-$$$$

Neptune's Hideaway

104180 Overseas Highway; Key Largo, FL 33037; tel: 305-451-0357.

These pink-and-white cottages sit on a private beach decorated with thatched umbrellas. Each unit contains two separate guest facilities, some with full kitchens, sitting rooms, and screened porches. Rooms are modern and cozy. Diving trips and jet-ski rentals are available. $$

Westin Beach Resort

97000 Overseas Highway; Key Largo, FL 33037; tel: 800-539-5274 or 305-852-5553.

This four-story complex has 200 rooms with private balconies overlooking a 12-acre indigenous hardwood forest. The resort features a private white-sand beach, a marina, trips to the Everglades, and daily snorkeling and scuba-diving tours of the reef. The complex includes three restaurants, two pools with coral rock waterfalls, tennis courts, Jacuzzi, sauna, and nature trails. $$$-$$$$

TOURS & OUTFITTERS

Caribbean Watersports Enviro-Tours

P.O. Box 781; Key Largo, FL 33037; tel: 800-223-6728 or 305-451-3595.

Naturalists lead rafting excursions into isolated regions of the Everglades and Florida Keys. Tours frequently encounter manatees, dolphins, bald eagles, and a wide variety of other birds. Itineraries are flexible.

Coral Reef Park Co.

P.O. Box 1560; Key Largo, FL 33037; tel: 305-451-1621.

The park's concessionaire offers daily snorkeling and scuba diving tours. Other services include a sailing-and-snorkeling catamaran excursion, a two-hour cruise in a glass-bottom boat, a dive shop, and boat rentals.

Florida Keys Dive Center

P.O. Box 391; Tavernier, FL 33070; tel: 800-433-8946 or 305-852-4599.

The center offers diving-lodging packages and all levels of instruction and certification at more than 100 sites in the upper Keys. Equipment rentals are available. Dive sites in the park include French Reef, noted for its beautiful staghorn coral and swim-through caves, and the popular Molasses Reef, with regal elkhorn and star coral.

Marine Resources Development Foundation

P.O. Box 787; Key Largo, FL 33037; tel: 800-741-1139.

The foundation offers two- to six-day MarineLab programs, involving interactive discussions with marine biologists, snorkeling in sea-grass beds, mangrove creeks, and coral reefs, and laboratory explorations of plankton and other tiny sea life. Longer programs are devoted to the ecology of coral reefs and the Everglades.

Silent World Dive Centers, Inc.

P.O. Box 2363; Key Largo, FL 33037; tel: 800-966-3483 or 305-451-3252.

Boats take a maximum of 15 passengers on diving tours of Key Largo National Marine Sanctuary and John Pennekamp State Park. Several multiday dive packages are available. The centers also offer rentals of diving equipment and underwater cameras, and all levels of diving instruction.

Excursions

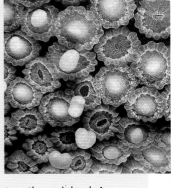

Biscayne National Park

P.O. Box 1369; Homestead, FL 33090-1369; tel: 305-230-7275.

Biscayne encompasses about 45 small islands at the northern tip of the Florida Keys, but most of the park is underwater. Here scuba divers and snorkelers can explore the northernmost coral reef in the hemisphere. Cruises aboard a glass-bottom boat depart from Convoy Point on the mainland. A seven-mile trail on Elliott Key is a good introduction to the plants and animals of subtropical hardwood jungle, including such threatened creatures as indigo snakes, crocodiles, and Schaus' swallowtail butterflies.

Dry Tortugas National Park

c/o Everglades National Park; 40001 State Road 9336; Homestead, FL 33034; tel: 305-247-6211.

Situated in the Gulf of Mexico about 68 miles from Key West, Dry Tortugas is one of the most remote and least-visited parks in the lower 48 states. The park protects 100 square miles of coral reefs and seven small islands, including Garden Key, site of a massive 19th-century fortification. Below the surface, the park is a rainbow world where staghorn and elk coral reach for sunlight, sea fans wave gracefully, sea turtles paddle lazily in the waves, and schools of brilliant blue tangs dart safely out of reach. There's an abundance of life above the surface, too. Dozens of bird species may be spotted in a single day. At times, more than 100,000 birds – mostly sooty terns and brown noddies – nest on the islands.

Virgin Islands National Park

6010 Estate Nazareth; St. Thomas, VI 00802; tel: 340-775-6238.

The park protects about two-thirds of St. John and the surrounding waters. Though small, the island has a surprising diversity of habitats, ranging from mangrove swamps and desertlike scrub to lush subtropical forests and fragile coral reefs. Diving tours can be arranged on nearby St. Thomas, or try diving directly from the beaches, considered to be some of the most beautiful in the Caribbean.

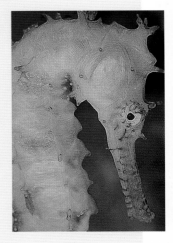

Boundary Waters
Canoe Area Wilderness
Minnesota

CHAPTER 10

Basswood River churns noisily over the lower falls and plummets into **Crooked Lake**. A calm, warm day lies heavily on the canoe country, blue skies punctuated with white comma clouds. Two canoeists stroke silently along near the edge of an emerald stand of horsetail reeds. Rounding a bend below the falls, they stop paddling, letting their canoe drift quietly. Forty yards ahead, and but a few yards from shore, stands a cow moose with her tawny twin calves, each miniature moose peering from beneath the protection of the mother's neck. The paddlers gently back away, giving the moose family space. Slowly the cow turns and sloshes onto the beach, then drips into the dark forest. Almost reluctantly, the calves turn to follow, seemingly as amazed at what they had just seen as are the couple in the canoe. ◆ Spotting wildlife ranks high on the list of hoped-for experiences of visitors to the Boundary Waters. No one will ever forget his or her first

Floating by canoe is the best way to spot the moose of the boreal forests and hear the music of the loon.

encounter with a moose, or discount the sight of a bald eagle wheeling overhead. **Boundary Waters Canoe Area Wilderness** – a 150-mile swath of boreal forest and bejeweled lakes on Minnesota's border with Ontario – is like few other natural areas. Protected by various laws since the mid-1920s, finally becoming a part of the National Wilderness Preservation System in 1978, it is America's only lake-land wilderness. More than a thousand lakes, hundreds of miles of creeks and rivers, and a million acres provide the best flat-water canoeing in the United States and a preserve for boreal forest and the species that inhabit it.

Canoeing offers ideal access to the remote areas of Boundary Waters; visitors can quietly approach and observe the preserve's wildlife.

Unlike wildlands in the West where vistas reach to the horizon, the Boundary Waters' lack of topographical relief and dense northern forest of conifers, birch, and aspen pose special challenges to those who would view its wildlife. The craggy landscape of Canadian shield granite and the thick boreal forest make overland hiking nearly impossible. For those intent upon finding wildlife in this lake-land wilderness, it should come as no surprise that water is the key. Nearly all nonwinter travel in this region is done by canoe from one of the dozens of Boundary Waters entry points. Like the native people who lived here for centuries, today's visitors traveling the wilderness by canoe carry their craft over the short trails, called portages, that link the myriad lakes. Outfitters in Ely, two hours north of Duluth, and in other locations adjoining Boundary Waters rent canoes and equipment, recommend routes, and offer guided trips.

Wildlife also use lakes and shorelines as travel corridors or as places to feed, resulting in frequent encounters. Among the animals most visitors hope to see are moose, which are fairly common and relatively easy to find, and timber wolves, also common but rarely seen. Other large mammals gracing canoe country are black bears, white-tailed deer, lynx, fishers, beavers, and otters.

Howls in the Night

The reclusive wolf is the glamour mammal of canoe country, but except in the winter when wolves are sometimes seen crossing the white, frozen lakes, you will rarely spot more than wolf tracks and droppings. Still, even this can be a thrill.

Beaches are good places to watch for wolf tracks, as are muddy areas on portage trails. Summer wolf droppings are seldom seen; dried winter scat can be found frequently once you know what you are looking for and where to search. In winter, wolves feed primarily on moose or deer, consuming the entire animal, hide and all. The indigestible hair is passed through the wolf and, when found months later on a summer day, looks like a twisted chunk of rope about five inches long and an inch thick. High ridges, which are frequent wolf travel routes, are the best places to find this telltale sign. They are also great places to pick blueberries.

The odds are much greater that you'll hear a wolf rather than see one, and the experience is exciting. While wolves may howl at any time, frequency increases an hour or so after complete darkness, and again just at sunrise, particularly on rainy days. Sometimes you can start the chorus yourself just by howling. Wolves up to two miles away may answer.

Moose, by contrast, are plentiful and among the most commonly spotted large mammals. The chances of seeing them are best near water. Slow waterways – especially those with lots of water lilies, a favorite moose food – are always excellent places to watch for these monster deer. Look for moose at dawn and dusk, and all day long on very hot midsummer days in shallow back bays where they can submerge themselves to stay cool and feed at the same time. Moose tend to like younger forests, and prime locations for spotting them are slow waterways near forests rejuvenated by fire. Any Forest Service office on the edge of the Boundary Waters can direct you toward areas that have recently experienced fire. Although you probably know that moose feed on aquatic plants during the summer, you may not be aware that white-tailed deer do the same on occasion. Woodland caribou once roamed the Boundary Waters region, but logging changed the habitat to favor the whitetail. Now that the Boundary Waters

is protected and the forest is maturing once again, there is talk of reintroducing caribou in the near future.

Beavers and Bear

Waterways near younger forest are also prime locations for spotting the ever-busy beaver. Telltale signs of these rodents can easily be found during your travels, from the mounded lodges of sticks along banks to the dams that frequently cross the creek on which you're traveling. Dams that are leaking or overflowing from recent rains are good places to park quietly at dusk and await the crew of beavers sent to make repairs. Sometimes a clump of green branches mysteriously moves across open water, as if

Coyotes (right) mate in the winter; young pups are ready to emerge from their dens after the spring thaw.

Lynx (opposite, above) have large, padded feet for moving across deep snow.

Red squirrels (opposite, below) are not shy about making their presence known. Listen for their persistent, staccato calls.

Beavers use their strong jaws and teeth to fell trees for their dams and lodges and to dine on the cambium layer of bark.

self-propelled. What you may not see is the beaver beneath that clump, dragging its fresh-cut dinner to its lodge. One certain way of determining whether a beaver lodge is occupied is to look for these branches stuck in the mud beneath the surface of the water near the lodge. As autumn nears, beavers feverishly jam branches of aspen, birch, and alder into the lake or pond floor where they will remain supple and tasty through the winter.

Other creatures are active at water's edge. Black ducks, ring-billed ducks, and common mergansers nest along the banks. Mink and fishers dart among the dark rocks, and otters travel on the banks or swim near them. Of all the canoe country's wildlife, otters are the most curious and joyful. They sometimes approach the quiet canoeist, emerging from the water like a furred periscope mere yards away. Clumps of shells from mussels or crayfish on shoreline rocks are signs that an otter or two enjoyed a seafood dinner.

This tangled northern forest is also home to one of North America's largest carnivores, the black bear. It is not unheard of to see a black bear swimming far out in a lake as it makes for a distant island campsite with the thought of raiding the camper's stores. Most encounters with black bears revolve around such incidents and happen most often near dawn in a daring raid. The black bear is rarely aggressive toward humans, but it displays clever problem-solving skills as it tries to figure a way to get at the freeze-dried lasagna. When

Taking to the Water

The canoe, the perfect craft for the Boundary Waters, is an excellent wildlife observation platform. Here are some tips for enjoying safe canoeing and productive wildlife watching:

● Pick a canoe made of a quiet material, such as wood or plastic, rather than metal. A canoe that silently slips through water helps you avoid spooking wildlife.

● Always wear a lifejacket. Canoes tip easily.

● Put cameras and binoculars in a watertight container, such as a dry bag.

● Travel quietly, as sound carries great distances over water. Try not to bang the canoe with the paddles, and keep your party size under four people.

● Always paddle near shore. Not only is this safer, you'll see far more wildlife.

● Choose seldom-traveled routes if you want to spot moose.

● Back your canoe away from a creature that begins to act nervous. Give animals their space.

● Obtain a permit from the Forest Service for canoeing Boundary Waters. Day-use permits are unlimited; overnight permits are limited.

Avoid spooking wildlife by paddling as quietly as possible. Wood or plastic canoes don't make as much noise as metal.

Gray wolves (opposite), also known as timber wolves, leave their tracks along rivers and on muddy trails.

you're canoeing during the day, carefully scan semi-open hillsides that may harbor blackberries, raspberries, or blueberries. All are favorite foods for black bears.

An Abundance of Birds

Whatever your Boundary Waters route, you will almost surely see bald eagles and ospreys, though rarely on the same lake, as bald eagles won't tolerate ospreys within their territory. The nests of each species are large and impressive, and made of sticks, with the eagle's typically one-third down from the treetop and the osprey's very nearly at the crown. Large Norway or white pines near the water's edge are favored nest sites, for both raptors depend upon fish as food. Sit quietly in your canoe at dawn within sight of a nest and you may be fortunate enough to watch an eagle or osprey deliver a fish to its mate and offspring. Listen and you'll hear the insistent cries of the young, which are constantly hungry, needing to fuel growth that takes them from hatchling to adult in just four months. Winter comes early in canoe country, and young eagles and ospreys must grow rapidly to be ready for the migration south.

Along with the other large soaring bird of the Boundary Waters – the turkey vulture – both the eagle and the osprey loft skyward on warm updrafts on blue summer days, drifting with nary a wingbeat. All three are nearly the same size and sometimes difficult to tell apart at a distance. Savvy birders will note that the vulture soars with its wings upward in a slight V, while eagles soar with wings nearly flat. The osprey cups its wings slightly, giving the impression of a flattened, upside-down W.

A trip to the Boundary Waters would be incomplete without the serenade of the common loon. These black-and-white, goose-sized water birds nest throughout canoe country and by the Fourth of July swim along trailed by their one or two slate-gray chicks. These are fish-eating birds, with a rapier bill from which few targeted fish can escape, and they are among the most vocal of this region's wildlife. Their haunting calls drift across the water at just about any hour of the day or night, but it is the evening serenade that most enchants you as the birds yodel across the dark, still waters.

Ospreys (left) nest near water, where they can find a supply of fish to feed their young.

Young loons (below) are able to swim when only a few days old, yet also like to be ferried by an adult.

Bald eagles (opposite, below) are adept at catching their own fish but also harry ospreys to release their catch.

River otters (opposite, above), active in winter, sometimes can be observed playfully scooting down snow slides.

With 27 species of wood warblers and a host of boreal specialists like flycatchers and the black-backed woodpecker, birding in the boreal forest of the Boundary Waters is very special. You will undoubtedly encounter the precocious gray jay, also known as the Canada jay, as it dares to glide noiselessly into picnic sites to make off with bits of bread or other morsels. The habit has earned this friendly jay the nickname of "camp robber." Two types of grouse – the common ruffed grouse and the rarer spruce grouse – are native to canoe country. Odds are that you will see these birds when walking quietly on portages. Spruce grouse are always darker and do not have the black-banded tail for which the ruffed grouse is known.

Woodpeckers of all sorts, from the common downy and hairy to the rarer pileated and even rarer black-backed, will treat you to their staccato wrapping, which can lead the quietly stalking birder to the bird. The great blue heron is common to this wilderness. When surprised, herons stretch out their long, slender neck and go absolutely rigid. Thus frozen, and with their gray coloration, they can be very difficult to spot. Watch for them in shallow waters and marshy areas where they wade along on their spindly legs looking for frogs and minnows. While you're there, keep your eye open for a richly brown bird about two feet long, also frozen with its pointy bill and head angled straight to the sky. This is the American bittern, a relative of the heron. Its call, a sort of a *boom ba-doom,* sometimes echoes from the marshes.

It is a magic place, the Boundary Waters, and only the gleaming waterways open its dark secrets to the visitor. The wildlife experiences are frequently intimate – and usually unexpected. For the very reason that there are few predictable places where wildlife can be observed, the experience is that much more special. When people and wildlife do come together, both are surprised. Like the calf moose reluctant to follow their mother into the forest, it may be questionable as to who is watching whom.

TRAVEL TIPS

DETAILS

When to Go

Cold temperatures and strong winds are always a possibility. Lakes remain prohibitively icy into May. Summer temperatures range from 80°F in the day to 30° at night. Fall temperatures range from the 50s to below freezing.

How to Get There

Duluth International Airport is about a two-hour drive from most park entry points. Car rentals are available at the airport.

Getting Around

Canoeing is the best way to see Boundary Waters, but the park also has 245 miles of trails, many of which can be reached by land. Most lakes do not permit motor-powered watercraft; others have horsepower limits.

Backcountry Travel

Permits are required for overnight trips or motorized day trips from May 1 to September 30. Quotas are enforced. To request a permit, call 800-745-3399 or write BWCAW Reservation Center; P.O. Box 450; Cumberland, MD 21501-0450. Self-issuing permits are required for day use or overnight trips from October 1 to April 30. To obtain a self-issuing permit, call the reservation center or visit any of the forest's main entry points. Camping is permitted only at U.S. Forest Service campsites that have steel fire grates and wilderness latrines or within designated primitive management areas.

Handicapped Access

All district ranger offices and some campgrounds are accessible. Wilderness Inquiry conducts canoe excursions for handicapped individuals. For information, call 800-728-0719 or write to Wilderness Inquiry; 1313 Fifth Street SE; Box 84; Minneapolis, MN 55414-1546.

INFORMATION

Superior National Forest

8901 Grand Avenue Place; Duluth, MN 55808; tel: 218-626-4300.

Grand Marais Chamber of Commerce

15 South Broadway; P.O. Box 1048; Grand Marais, MN 55604-1048; 888-922-2221 or 218-387-1400.

Minnesota Tourism

500 Metro Square; 1217 Place; St. Paul, MN 55101; tel: 800-657-3700 or 612-296-5029.

CAMPING

There are 27 developed campgrounds in Superior National Forest; most are operated by concessionaires. About half of the campgrounds accept reservations. To make a reservation, call 800-280-2267.

LODGING

PRICE GUIDE – double occupancy

$ = up to $49 $$ = $50 - $99

$$$ = $100 - $149 $$$$ = $150 +

Ludlow's Island

P.O. Box 1146B; Cook, MN 55723; tel: 218-666-5407.

Eighteen cabins on Vermilion Lake, between Boundary Waters and Voyageurs National Park, offer quiet and contemporary ease. No two cabins are alike on this island, where finely crafted buildings sport cathedral ceilings, massive beams, striking pine interiors, and generous decks and screened porches. The cabins have one to five bedrooms, at least two bathrooms, a kitchen, and a fireplace; some have a sauna or whirlpool, washer and dryer, and private beach. Amenities include a lodge, grocery store, tennis and racquetball courts, an exercise room, a recreation room, sailing, camping, and boat rentals. Open May to October. $$$$

Lund's Motel and Cottages

919 West U.S. Highway 61; P.O. Box 126; Grand Marais, MN 55604-0126; tel: 218-387-2155 or 218-387-1704.

Situated on the "Scandinavian Riviera," Lund's offers modern housekeeping cottages and large motel units in a secluded, wooded setting. Family-owned since 1937, Lund's is across the road from the local indoor pool, sauna, and whirlpool. Rooms have queen-sized beds and fireplaces. $$

MacArthur House Bed-and-Breakfast

520 West 2nd Street; P.O. Box 1270; Grand Marais, MN 55604-1270; tel: 800-792-1840 or 218-387-1840.

Located a half-block from the Gunflint Trail and two blocks from the harbor in Grand Marais, this inn offers five spacious guest rooms with cathedral ceilings and private baths. The new three-story house features art in North Shore and Alaskan style and a fireplace in the living room. Amenities include Jacuzzi and ski-tuning room. $$

Nelson's Resort

7632 Nelson Road; Crane Lake, MN 55725; tel: 800-433-0743 or 218-993-2295.

This resort in Superior National Forest has 29 cabins on more than a mile of lakeshore. Log and pine cabins are rustic but comfortable. Each cabin has a private bath; two-bedroom units have two baths. Some cabins have kitchens, decks, and sitting areas. The hand-hewn log lodge has three fireplaces and a dining room. Rentals of bicycles, canoes, and motorboats are available. Inquire about fishing guides. $$–$$$

Nelson's Travelers Rest Cabins and Motel

West U.S. Highway 61, Lakeside; P.O. Box 634; Grand Marais, MN 55604-0634; tel: 800-249-1285 or 218-387-1464.

This quiet in-town resort has nine shaded log cabins and two motel units. Large one- and two-bedroom cabins offer modern comfort, with a private bath and fireplace; some have a deck, queen-sized bed, fully equipped kitchen, and a view of Lake Superior. Motel units sleep two to four people; each has a deck, microwave, and refrigerator. Amenities include playground, picnic tables, lawn chairs, barbecue grills, outdoor fire pit, and free firewood. Open May to October. $–$$

Northern Lights Resort

12723 Northern Lights Road; Ray, MN 56669; tel: 218-875-2591.

This 16-acre resort sits on Lake Kabetogama in Voyageurs National Park, just a few miles from the Canadian border. Housekeeping cabins, situated at water's edge, have knotty-pine paneling, queen-sized beds, modern kitchens, and decks with barbecue grills. Two cabins offer two full baths. A log lodge has a game room and stone fireplace. Also on the premises are a small grocery store, a playground, water sports equipment, and a sand beach. Cabins rented on a weekly basis. $$–$$$

TOURS & OUTFITTERS

Dozens of outfitters and guides work in the region. To request a list of those licensed to operate in Boundary Waters, call Superior National Forest at 218-626-4300.

Excursions

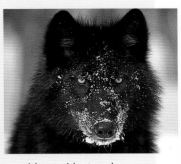

Agassiz National Wildlife Refuge

Route 1, Box 74; Middle River, MN 56737; tel: 218-449-4115.

Moose and wolves range this 100-square-mile marsh and woodland refuge. Agassiz is the only national wildlife refuge in the lower 48 states with a resident pack of gray wolves. Visitors rarely catch sight of these creatures, although they may hear them howling or see their sign. Scarlet-crowned sandhill cranes nest here in spring and congregate by the thousands in fall. The refuge is a breeding ground for five species of grebes and more than a dozen species of ducks. Other birds include the black tern, white pelican, and various species of bitterns and grouse.

Isle Royale National Park

800 East Lakeshore Drive; Houghton, MI 49931-1869; tel: 906-482-0984.

This island in Lake Superior is reached only by boat or floatplane. The park, which has one lodge and no roads, is best suited for those courting isolation. Throughout the forest of white spruce and balsam fir are red foxes, beavers, red squirrels, moose, and gray wolves. Loons, among the island's 200 bird species, epitomize this rugged park. Their odd, mournful cries are often heard across coves and inland lakes.

Voyageurs National Park

3131 Highway 53; International Falls, MN 56649-8904; tel: 218-283-9821.

The park is named for the French traders, or *voyageurs*, who paddled these waters in the late 18th century. Receding glaciers left behind a maze of tiny islands and waterways, and the best way to see them is by boat. Canoes, motorboats, and houseboats may be rented at several entry points. Tours led by a park naturalist are also available. Beavers, bald eagles, river otters, moose, and loons are regularly seen. Present but less visible are gray wolves and black bears.

Buffalo
National River
Arkansas

CHAPTER **11**

An October morning is well along by the time the sun tops a Buffalo River bluff and sends golden light flooding into this Ozark Mountain valley in Arkansas. On the slopes above the placid stream, the leaves of oaks, hickories, sweet gums, and maples seem to glow in every possible shade of yellow, orange, and red. ◆ A bald eagle, just arrived from its nesting grounds in Ontario, leaves its perch in a sycamore and heads upriver on deep, powerful wingbeats. As it rounds a bend, it passes a lone fisherman, close enough that the man can hear the wind in the flight feathers and see the gleam in the bird's eye. "Lordy, lordy," he says softly, with something like reverence in his voice. ◆ Around the next curve, the eagle startles a great blue heron standing motionless on the riverbank. The gangly wader knows instinctively that the raptor is no threat, but it takes flight with awkward flaps and a guttural squawk. On the opposite bank, the eagle spots an osprey resting in a tall white oak, and is eyed suspiciously in return. Bald eagles are known to steal ospreys' hard-won prey, but having just finished its morning meal, the "fish hawk" has nothing to take. ◆ Now it's the eagle's turn. Spotting movement below, it wheels in the air and folds its wings, dropping to just inches above the river. At the last moment, it extends its massive talons and snatches a luck-less fish from the limpid water. Climbing slowly, the hunter flies toward a nearby limb, only to shy away on its approach. A black bear is snuffling along the ground just below, searching for acorns and hickory nuts – aware that winter is on the way, and with it harder times.

Along a watery path through the Ozarks, ospreys dive for fish, black bears forage the shore, and bats hunt insects at dusk.

Northern cardinals nest along the river. The female, though drabber than the brilliant male, can easily be spotted bringing food to the young.

The eagle glides off to another perch, to feed in peace. Full and content, it will spend the rest of the morning loafing, watching the river run, in a scene as timeless as these weathered hills.

Ribbon of Wilderness

The **Buffalo River** is born in the rugged **Ozarks** of northwestern Arkansas, where dozens of tiny creeks tumble down rocky "hollers" and join to create one of America's most beautiful and unspoiled streams. Generations of canoeists, anglers, swimmers, and campers have delighted in the Buffalo, which alternately rushes in churning rapids and flows lazily in long pools on its way to a confluence with the **White River**, 150 miles from its source.

It's only thanks to a corps of dedicated conservationists that the Buffalo remains wild and free today. Like many other Ozark streams, the river was threatened with death by dam in the 1950s and 1960s. But in 1972, after a long and bitter fight, the Buffalo was designated as America's first national river, administered by the National Park Service from headquarters in nearby Harrison.

In spring, the upper Buffalo, from the tiny hamlet of **Ponca** down to the Highway 7 bridge north of **Jasper**, attracts thousands of canoeists with its challenging whitewater and imposing limestone bluffs, some of which tower 500 feet over the water. While the ride is thrilling, it's hard to see much wildlife when you're trying to pick the right line through surging rapids and avoid obstacles like the infamous **Gray Rock**, upsetter of countless canoes. Better to stop occasionally on a gravel bar beside a quiet pool, where you may hear such beautiful songsters as a hooded warbler or scarlet tanager from the streamside trees, or see a green heron fishing on the bank. Or, if you're truly lucky, catch a glimpse of a mink scurrying across the rocks, perhaps with a frog or small fish in its

Woodchucks (left) are daytime foragers that seek out fresh, tender plants like these dandelions.

Ospreys (opposite, above) dive into the water to snag fish with their talons, then carry them away head first.

Armadillos (opposite, below) are difficult to confuse with any other North American animal; they roam the areas beyond the river, especially at dawn and dusk.

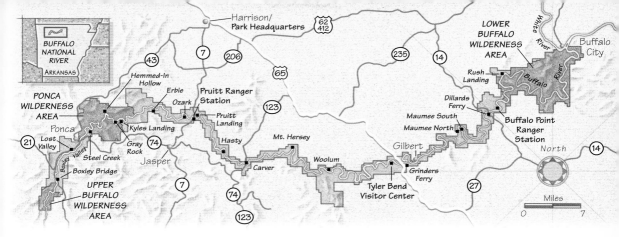

mouth. For another break from paddling, take the short walk to **Hemmed-In Hollow**, where spring rains feed the highest waterfall between the Appalachians and the Rockies, a 175-foot ribbon of spray dropping into a dramatic rock amphitheater.

On the lower Buffalo, the rapids dwindle to a few relatively gentle riffles, easily negotiated even by beginning floaters. Here, canoeing is less challenge than sheer relaxation, the river and its wooded banks a world apart from freeways and malls. Lean back and you may see a red-shouldered hawk soaring overhead, its high scream evoking the wildness of the landscape. Drift quietly round a bend and you may surprise a white-tailed deer doe and fawn drinking, silently slipping back into the forest as you approach. Look down and you could spot a smallmouth bass in the Buffalo's crystal-clear water. Much of the pristine river habitat of this sought-after game fish has been drowned beneath reservoirs or lost to pollution, but it still thrives in the Buffalo.

Those who have no desire to float will still find plenty of opportunity for wildlife observation along the Buffalo. Miles of hiking trails parallel the river and some of its tributaries, and three wilderness areas offer backcountry solitude. White-tailed deer are the most common large mammal in the region, but black bears have multiplied in recent years after near elimination

in the 1950s. The most likely bear sighting will be a view of a furry rump as the bruin runs away, though hikers should take the standard precautions.

Yet one more big critter roams the Buffalo country, and a sighting could well shock the unprepared. The Arkansas Game and Fish Commission has reintroduced elk here, where they last roamed more than a century and a half ago. Since the first Buffalo River elk were released in 1981, the population has done well enough that a limited hunting season was announced in 1998. For the best chance to see these magnificent animals, visit the **Boxley Valley** on the upper river and the **Steel Creek** and **Erbie** areas farther downstream.

Sunrise and Sunset Performances

As is so often true, dawn and dusk are by far the best times for wildlife watching on the Buffalo. In late April and early May, sunrise brings a chorus of birdsong from both residents and migrants. The lovely fluting song

Young gray foxes (left) develop hunting skills through play before they need to find their own food.

The spiny orb weaver (below) carries its own protection from enemies and often builds its webs high in trees.

Quiet creeks and pools (right) along the river are good places to watch for wildlife.

of the wood thrush echoes through the woods, and the stunning pileated woodpecker drums on a tree trunk and shouts its raucous laughing call. Along the river a few cerulean warblers prepare to nest, singing their buzzy songs from the tops of sycamores. This tiny bluish bird prefers large tracts of mature hardwood forest for its home, a habitat increasingly scarce. The steady decline in the cerulean's population has ornithologists concerned about its future. Its cousin, the northern parula, is far more common, though just as difficult to catch sight of in its inevitable treetop perch.

In open areas, early risers might come upon a male wild turkey in courtship display, but a glimpse of this wary species is more likely later in the year when females are shepherding their young through the woods. If the turkey brood happened to cross paths with a roving gray fox, it could well lose a young bird or two, but the pretty little fox, almost catlike in its ability to climb trees, is much more likely to dine on one of the cottontail rabbits ubiquitous in every scrubby field and wood-

land opening. Cottontails are prey, too, for the secretive bobcat, which only the most fortunate dawn trail-walker will see as it completes its nocturnal rounds.

Night Life

As evening comes, the night walkers and fliers make their appearance. A raccoon waddles to the riverbank, using its agile hands to grasp a crayfish in the shallow water. Bats flit overhead, using their amazing echolocation ability to home in on insects. A barred owl shouts its familiar *Who cooks for you? Who cooks for you-all?* call – or, just possibly, gives voice to its other vocalization: a series of blood-chilling, eerily human shrieks and screams that few people have heard. Eyes open wide and neck hairs rise with thoughts of gory murder, until someone says, "Oh, it's just an ol' hoot owl," and everyone laughs with relief.

The river soothes the way to sleep, but unlucky is the camper with a chuck-will's-widow or whip-poor-will for a neighbor. These two closely related night birds are

named for their piercing calls – notes they're capable of sounding hundreds of times in succession. More than one camper has risen in the middle of the night to shout in frustration at a disembodied voice resounding through the woods like an alarm clock that just won't shut off.

But sleep does come, perhaps to the whis- tled trill of an eastern screech-owl, blending with the noises of countless insects in every tree and shrub. Of the campers, some may have sore arms and backs from paddling, others sore legs from hiking. But all will have something new in their dreams: memories of the beautiful, free-flowing Buffalo, and of the wild creatures that make it their home.

TRAVEL TIPS

DETAILS

When to Go

Summer is hot and humid, with temperatures often in excess of 90°F. Spring and fall are pleasant, with average temperatures in the 50s and 60s. Thunderstorms, which produce good floating conditions, are common in spring and summer. Winter tends to be mild, with occasional light snow. Floating season on the upper river begins in the end of February, and in May on the middle and lower stretches of the river. Low water levels sometimes make float trips difficult from August to the beginning of October.

How to Get There

The nearest major airports are in St. Louis and Kansas City, Missouri, and Memphis, Tennessee. Inquire about connections to Little Rock and Fayetteville Municipal Airports in Arkansas or Springfield-Branson Regional Airport in Missouri.

Getting Around

Park headquarters are in Harrison, 80 miles from Fayetteville, Arkansas, and Springfield, Missouri, and 150 miles from Little Rock, Arkansas. Car rentals are available at the airports.

Handicapped Access

The visitor center, some trails (with assistance), and four campgrounds are wheelchair accessible.

INFORMATION

Buffalo National River
402 North Walnut, Suite 136; Harrison, AR 72601; tel: 870-741-5443.

Arkansas State Tourism
1 Capital Mall; Little Rock, AR 72201; tel: 800-628-8725 or 501-682-1088.

Harrison Chamber of Commerce
621 East Rush; Harrison, AR 72602; tel: 800-880-6265 or 870-741-2659.

CAMPING

The park's 14 campgrounds are accessible by car and available on a first-come, first-served basis.

LODGING

PRICE GUIDE – double occupancy

$ = up to $49 $$ = $50-$99
$$$ = $100-$149 $$$$ = $150+

Brambly Hedge Cottage
HCR 31, Box 39; Jasper, AR 72641; tel: 800-272-6259 or 870-446-5849.

Built around a 19th-century log cabin, this Cotswold-style cottage is set atop Sloan Mountain, overlooking the Little Buffalo and Buffalo National Rivers. A relaxing deck provides a grand view of the valley. Three guest rooms offer French country decor and private baths. A converted barn has living room, fireplace, two bedrooms, and private bath. Amenities include evening dessert and beverages, and a fireplace in the living room. $$–$$$

Buffalo Point Concession
Buffalo National River; HCR 66, Box 388; Yellville, AR 72687; tel: 870-449-6206.

The only lodging within the park has five rustic and eight modern cabins. Built in the 1930s, the rustic cabins have handcrafted furniture, fireplaces, and screened porches. Two of the eight duplex cabins afford a view of the river. All cabins have private baths, living rooms, and kitchenettes. $$

Little Switzerland
P.O. Box 502; Jasper, AR 72641; tel: 800-510-0691 or 870-446-2693.

Homey accommodations on 80 mountain acres laced with hiking trails. Private two-story cabins have queen-sized and double beds, full bath, balcony, refrigerator, and coffee maker. $–$$

Lost Spur Guest Ranch
4648 Lost Spur Road; Harrison, AR 72601; tel: 800-774-2414 or 870-743-7787.

Six cedar log cabins overlook the Ozark Mountains and Crooked Creek. Roomy lodgings are decorated in rustic western style. All have private baths, porches, and queen-sized and twin beds; some have two bedrooms and a kitchenette. Prices from April to October include cabin, meals, and dude ranch. Horseback riding, chuck wagon cookouts, canoeing, and hayrides are available. $$–$$$

Ozark Mountain Cabins
P.O. Box 509; Jasper, AR 72641; tel: 800-377-4805 or 870-446-2229.

These modern log cabins are set on a forested mountainside bordered by Buffalo National River and Ozark National Forest. Each cabin has a living room, stone fireplace, fully equipped kitchen, private bedroom with a queen-sized bed, loft with two beds, and covered porch with swing. $$

Wild Bill's Outfitter
HCR 66, Box 380; Yellville, AR 72687; tel: 800-554-8657 or 870-449-6235.

These modern cabins are perched on a scenic ridge in a grove of pine and oak trees. Each has a master bedroom with king-sized bed, loft bedroom with twin beds, living room with fireplace, kitchen, and front porch with swing. Some cabins have a two-person Jacuzzi. A grocery store and deli are on the premises. $$

TOURS & OUTFITTERS

Buffalo Outdoor Center
P.O. Box 1; Ponca, AR 72670; tel: 800-221-5514.

Canoe, raft, and mountain-bike rentals, horseback riding, and hot-air ballooning. Located 25 miles from park headquarters.

Buffalo River Outfitters
Route 1, Box 56; St. Joe, AR 72675; tel: 800-582-2244 or 870-439-2244.

Canoe and raft rentals, horseback riding, and guided trips in the river's middle region.

Lost Valley Canoe and Lodging
Ponca, AR 72670; tel: 870-861-5522.

Half-day to 10-day float trips on the upper Buffalo River. Paddlers explore dramatic bluffs and caves and enjoy prime views of the highest waterfall between the Appalachian and Rocky Mountains. Canoes and whitewater rafts are available.

Ozark Ecotours
P.O. Box 513; Jasper, AR 72641; tel: 870-446-5898.

Day or overnight hiking, back-packing, camping, canoeing, nature awareness trips, and horseback riding tours.

Wild Bill's Outfitter
HCR 66, Box 380; Yellville, AR 72687; tel: 800-554-8657 or 870-449-6235.

Year-round canoe rental by the hour, day, or week. Rental includes shuttle service to and from any point on the river. Rafts, kayaks, johnboats, and guide services are also available.

Excursions

Holla Bend National Wildlife Refuge
P.O. Box 1043; Russellville, AR 72801; tel: 501-968-2800.

Situated between the Ozark and Ouachita Mountains, this bend in the Arkansas River is a favorite of birders. Bald and golden eagles gather here each fall and winter along with tens of thousands of migratory waterfowl. Northern harriers and kestrels are frequently seen scouting for a meal in winter. There are also white-tailed deer, armadillos, beavers, coyotes, and black bears.

Ouachita National Forest
P.O. Box 1270; Hot Springs, AR 71902; tel: 501-321-5202.

The forest encompasses nearly two million acres of woodland and seven wilderness areas, but out-of-staters don't seem to know about it. Observant visitors will find richly varied wildlife – black bears, bobcats, white-tailed deer, river otters, red-cockaded woodpeckers, cerulean warblers, armadillos, and timber rattlesnakes. The forest also has one of the country's richest amphibian habitats, with many species of salamanders and frogs.

Ozark National Forest
605 West Main Street; Russellville, AR 72801-3614; tel: 501-968-2354.

Primarily located in northwestern Arkansas, the park's 1.2 million acres are rich with oak, hickory, and pine. Ranging among the 500 species of plants are black bears, turkeys, elk, white-tailed deer, and a variety of neotropical migratory birds. Several species of the rare Diana butterfly are attracted to Magazine Mountain, whose 2,753-foot peak is the tallest in the state. Several nationally designated wild and scenic rivers, including the Buffalo River, flow through the forest.

Platte River
Nebraska

CHAPTER 12

Before sunrise on a frigid March morning, a dozen strangers gather at a roadside gas station in south-central Nebraska. Bundled in down parkas and layers of long underwear, they huddle around cups of steaming hot coffee. When everyone is assembled, the caravan heads down a country road and parks beside a cornfield. The people spill out of their vehicles and listen intently as the leader whispers instructions. They've come here to view sandhill cranes, but no one is prepared for the sight they are about to see. ◆ The members of the group stumble across the stubbled frozen field, aiming for a line of trees along the river. Without a word, everyone crowds into a small wooden building, a blind from which the cranes will be visible. The blind is on the north side of the **Platte River**, at the **Lillian Annette Rowe Sanctuary**. Owned by the National Audubon Society, the sanctuary sponsors these expeditions each spring, so that visitors can experience the sights and sounds of one of

Migrating sandhill cranes, by the thousands, fill the Platte River valley with their fluttering wingbeats, trumpeting calls, and courtship dances.

North America's most spectacular wildlife events – half a million sandhill cranes congregating at this stopover on the Platte. ◆ Peering out of tiny portholes in the walls, the observers see cranes standing statue-still in the middle of the shallow stream, only a couple hundred yards away. The dim light of dawn allows only a subtle hint of the birds. They coo softly like mothers talking to their babies. No one makes a sound, lest the cranes be roused prematurely from their night roost. With sunrise, details of the birds' shapes and colors emerge, as if on a photograph in a developing tray: long necks, slender legs, gray plumage, vibrant red heads, three-foot-tall birds

Sandhill cranes start arriving in February. Several weeks later, the flocks number hundreds of thousands as the birds assemble along the river to prepare for their northward migration.

standing shoulder to shoulder as far as the eye can see.

As the hidden watchers look on, the sandhills start to grow restless. They shift from one spindly leg to another, flap their ungainly wings, and preen their feathers. For about an hour, by ones and twos, then threes and fours, they rise off the sandbar into the gray sky. The sound swells until a bald eagle flies over, and the entire flock of several thousand birds ascends in one instant. The cranes' frenzied calls are like applause filling a huge auditorium. It is a primal scene, far beyond the reach of human memory.

Carbo-Loading and Courting

Each spring about 80 percent of the world's sandhill crane population gathers along 75 miles of the "Big Bend" of the Platte, between the towns of **Overton** and **Grand Island**. Here the cranes pause for a few weeks to rest and feed in the course of a momentous journey from south to north. Three populations, or subspecies – mostly lesser sandhill cranes, with a few greater and Canadian sandhills mixed in – follow the Central Flyway from wintering grounds in the Southwest states and Mexico to

Canada, Alaska, even Siberia. There they breed, nest, and raise their young. But it is here, in Nebraska, that they perform their courtship dances, the graceful leaps and bows that attract a lifelong mate. The birds' arrival time in Nebraska depends on temperatures – if it stays very cold, they may delay their arrival by a month or so. Generally, though, the cranes reach peak numbers by the third week in March. Most are gone by mid-April.

To fortify themselves for the remaining 4,000- to 5,000-mile migration, the cranes spend their days fattening up on waste corn in surrounding fields, increasing their body weight up to 10 percent. "The longer they stay and fatter they get," says Rowe Sanctuary director Paul Tebbel, "the later they take off in the morning." When they've had enough carbo-loading, the birds will leave on the next leg of their marathon.

Each evening, as the sun goes down, observers gather on bridges across the Platte. The spectacle of the cranes returning to their night roosts is not as dramatic as the morning departure, but it's still impressive. Wave upon wave of flocks come in for a landing. Through binoculars, they look like clouds of black smoke in the sky. They circle in, and, when the glide slope is right, lower their legs and float down to safety on a sandbar. The only sound of the birds overhead is the *whoosh* of their wingbeats on the silky night air.

Local newspapers run weekend features about this annual ornithological rite. A gaggle of experienced bird-watchers, plenty of neophytes, and interested locals brave brisk

Snow geese (left) winter in south-central Nebraska, congregating in large, vociferous flocks.

Sandhill cranes (opposite) leave the Platte in the morning to feed in the surrounding fields, then return to the river to spend the night.

temperatures to share in this wondrous phenomenon. Some come on their own, others on guided tours. At the Rowe Sanctuary, people can visit one of several blinds. By special arrangement, two people can position themselves for the morning drama by renting a blind for the night. Sanctuary managers advise these hardy crane watchers to bring not one but two sleeping bags each. The blinds are not heated, and subzero cold is not uncommon here in March.

Essential Waterfowl Habitat

In her last will and testament, New Jersey psychologist Lillian Annette Rowe directed that her inheritance go "to establish a bird sanctuary anywhere in the nation." Two years after her death in 1971, the Audubon Society fulfilled her wish by buying 782 acres along the middle channel of the Platte River at the town of **Gibbon**. The society knew this was a critical stretch of habitat for sandhill cranes and other waterfowl. Now doubled in size, the sanctuary protects one of the last nearly pristine stretches of the river to offer what cranes need: bare sandbars, shallow water and sloughs, and a broad channel largely unclogged by vegetation.

The islands, sandbars, and shores within Rowe Sanctuary are cleared each year of trees and brush, to keep the river channel open and attractive not just to cranes but to hundreds of thousands of waterfowl. Represented here are many kinds of ducks, including northern pintails, common mergansers, and blue-winged teals, as well as Canada geese, snow geese, and greater white-fronted geese. Nearly two million snow geese have been counted on the wetlands of the Rainwater Basin south of the Platte.

A few endangered whooping cranes occasionally gather with migrating sandhills. In 1997, an adult and juvenile whooper stayed all of March in central Nebraska. In late April, three adult whoopers created lots of excitement by spending four nights on the Platte, just outside the back door of the Rowe Sanctuary visitor center. Least terns, piping plovers, and Eskimo curlews are other species of concern that depend on the Platte River habitat. For all these reasons, many people have joined forces to ensure the survival of this river and the magnificent birds that need it so much.

TRAVEL TIPS

DETAILS

When to Go

Sandhill cranes begin arriving in mid-February, with peak numbers occurring in mid-March. Most of the cranes leave in the first week of April. Weather in March is unpredictable. Temperatures often range from the 50s to –10°F, and heavy snow is possible.

How to Get There

Gibbon is three hours from Epply International Airport in Omaha, two hours from Lincoln Municipal Airport in Lincoln, one hour from Central Nebraska Regional Airport in Grand Island, and 20 minutes from Kearney Municipal Airport in Kearney. Each airport is serviced by at least one major airline.

Getting Around

Car rentals and buses are available at the airports. An automobile is essential for travel in the area.

Handicapped Access

Handicapped assistance at the Lillian Annette Rowe Sanctuary and the Crane Meadows Nature Center may be arranged in advance.

INFORMATION

Nebraska Travel and Tourism
P.O. Box 98907; Lincoln, NE 68509-8907; tel: 800-228-4307 or 402-471-3796.

Kearney Visitors Bureau
P.O. Box 607; Kearney, NE 68848; tel: 800-227-8340 or 308-237-3101.

Lillian Annette Rowe Sanctuary
44450 Elm Island Road; Gibbon, NE 68840; tel: 308-468-5282.

CAMPING

Fort Kearny State Historical Park and Recreation Area
1020 V Road; Kearney, NE 68847; tel: 308-865-5305.

Sandhill cranes gather here in spring. The recreation area, eight miles from Rowe Sanctuary, has 110 campsites, a hiking trail, and a walking bridge over the Platte River, excellent for watching cranes. The visitor center conducts nature programs during crane season.

Windmill State Recreation Area
P.O. Box 427; 2625 Lowell Road; Gibbon, NE 68840; tel: 308-468-5700.

Situated four miles from the Rowe Sanctuary, this recreation area has three antique windmills and year-round camping facilities near small lakes.

LODGING

PRICE GUIDE – double occupancy
$ = up to $49 $$ = $50-$99
$$$ = $100-$149 $$$$ = $150+

Aunt Betty's Bed-and-Breakfast
804 Grand View Avenue; Ravenna, NE 68869; tel: 800-632-9114 or 308-452-3739.

This striking in-town Victorian was built as a boardinghouse in 1907 and recently restored. Four guest rooms share two baths. Each room has its own style of decoration; all rooms are furnished with antiques. $–$$.

Country Inn and Suites by Carlson
105 Talmadge Street; Kearney, NE 68847; tel: 800-456-4000 or 308-236-7500.

Three miles from Kearney Municipal Airport, this recently renovated inn has 56 rooms and 17 suites. Standard rooms have two queen-sized beds and private baths; suites have king-sized beds, sitting areas, and private baths, some with whirlpools. A restaurant, indoor pool, whirlpool, and health club are on the premises. $$

Grandma's Victorian Inn
1826 West 3rd Street; Hastings, NE 68901; tel: 402-462-2013.

Built in 1886, this restored Victorian house is situated in a quiet neighborhood less than two miles from the Hastings Museum. Five elegant guest rooms offer queen-sized beds, antique furniture, and private baths with clawfoot tubs. Reproduction Victorian wallpaper adorns the house throughout. $$

Kirschke House
1124 West 3rd Street; Grand Island, NE 68801; tel: 800-381-6851 or 303-381-6851.

An ivy-covered brick Victorian with stained-glass windows, this bed-and-breakfast is located near downtown Grand Island, 45 minutes from the Lillian Annette Rowe Sanctuary. Offering turn-of-the-century country decor, the house has four rooms with shared baths, one with private bath, and a two-level bridal suite with private bath, fireplace, four-poster queen-sized bed, veranda, refrigerator, and microwave. An old brick washhouse shelters a hot tub. $$–$$$

Memories Bed-and-Breakfast
900 North Washington; Lexington, NE 68850; tel: 308-324-3290.

You'll find a white picket fence and shade trees in the front yard of this Victorian inn and a gazebo, cobblestone patio, and two lovely porches in back. The house, built in 1903 and decorated in turn-of-the-century style, has four large guest rooms, three with queen-sized beds, two with private baths. Amenities include an attached antique shop and a "bicycle car." $–$$

TOURS & OUTFITTERS

Crane Meadows Nature Center

9325 South Alda Road; Wood River, NE 68883; tel: 308-382-1820.

Naturalists lead sunrise and afternoon tours during crane season (early March to mid-April). The center, which features indoor wildlife displays, also conducts hundreds of nature programs each year. Located on 258 acres of prairie, wet meadow, and riparian forest between two channels of the Platte River, this year-round educational site is the home of regal fritillary butterflies, smooth green snakes, white-tailed deer, coyotes, and badgers. More than 220 bird species have been seen at the nature center.

Lillian Annette Rowe Sanctuary

44450 Elm Island Road; Gibbon, NE 68840; tel: 308-468-5282.

Sanctuary guides lead trips to wooden viewing blinds in March. Blinds hold from 10 to 35 people and are available by reservation. Those wishing to come on weekends should make reservations three months in advance.

MUSEUMS

Hastings Museum

1330 North Burlington; Hastings, NE 68902; tel: 402-461-2399.

Three floors of nature and history exhibits. A renowned collection of mounted specimens features North American birds, with special emphasis on Nebraskan species. Also includes the world's largest display of whooping cranes in simulated habitat. The Lied IMAX Theater shows films daily on a five-story screen.

Excursions

Fort Niobrara National Wildlife Refuge

HC14, Box 67; Valentine, NE 69201; tel: 402-376-3789.

Elk and bison herds graze on rolling prairie and woodlands much as they have for hundreds of years, dropping calves in May and engaging in curious mating rituals in September. The refuge's 19,130 acres, bisected by the beautiful Niobrara River, also sustain coyotes, porcupines, white-tailed deer, mule deer, prairie dogs, and Texas longhorn cattle, a reminder of 19th-century cattle drives. Exceptional birding opportunities offer a chance to see sharp-tailed grouse, prairie chickens, wild turkeys, and a variety of raptors, including burrowing owls, which reside in abandoned prairie dog burrows.

Pawnee National Grassland

660 O Street; Greeley, CO 80631; tel: 970-353-5004.

This 193,060-acre preserve in northeastern Colorado is one of the few remaining expanses of short-grass prairie. Pawnee, now interspersed with some cultivated land, was once the grazing land of countless deer, antelope, elk, and bison. The elk and bison are gone, but 9,000 head of cattle now graze the allotments here, along with deer and pronghorn. The grassland supports more than 275 species of birds. Prairie chickens are particularly interesting. Males produce a low booming sound with brightly colored air sacs on their necks during courtship displays in spring.

Tallgrass Prairie Preserve

P.O. Box 458; Pawhuska, OK 74056; tel: 918-287-4803.

The finest reminder of what the tallgrass prairie – once 142 million acres – looked like, this preserve is home to 300 bird species and 80 types of mammals. One species – bison – commands singular attention. Bison were reintroduced in 1993, and the herd continues to grow. Coyotes, white-tailed deer, and armadillos are commonly seen; bobcats, beavers, and badgers are more elusive. Wildflowers blanket the preserve in spring.

Dakota Prairie
South Dakota

CHAPTER **13**

Testy during the rut, male bison roll in their dusty wallows and growl their displeasure at being disturbed by making a curiously liquid sound, low-pitched and gurgly, like water spilling over a steep ledge. At midday, with the late August sun bearing down, their dark, shaggy heads smolder like lumps of slow-burning coal. They move with ponderous dignity, heads swaying, their massive weight balanced on disproportionately slim forelegs. When you stand downwind from them, the stench wafting off their matted hides rakes through your nostrils like a wire brush.　◆　Encounters with bison – at a safe distance for you to get a good whiff – are one of the many experiences that make the South Dakota prairie such a special place. In spring and summer, hundreds of the legendary animals ramble in good-sized herds across the open grasslands of **Bison, pronghorn, elk, and deer** the southeast section of the Black Hills, taking **roam the vast prairies** their time, showing themselves off, **and forests of two adjoining** parading with a kind of haughty swagger. **parks in the Black Hills.** From numbers in excess of 40 million 150 years ago, the animals by 1890 had been reduced in North America to less than a thousand. Today, thanks to prudent conservation efforts, herds flourish at two contiguous parks. About 350 bison rove free in 28,295-acre **Wind Cave National Park**, known also for protecting one of the longest limestone caves in the world. **Custer State Park**, well over twice the size of the national park, supports a herd that, at its peak, numbers 1,500.　◆　Animals, big and small, have long been part of the Black Hills mystique. A Lakota Sioux legend describes how all the animals in the world raced around a huge flat circle to determine whether animals would eat humans or humans would eat animals. Before

Bison weigh up to 2,000 pounds, gallop at speeds reaching 35 miles per hour, and attack with their horns and large heads. Early observers considered them as dangerous as grizzly bears.

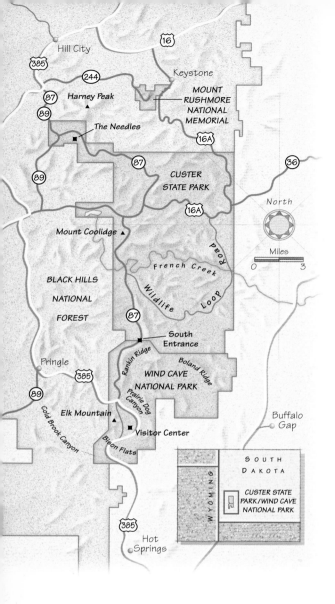

and mammals from voles, gophers, and chipmunks to such megafauna as bison, elk, and pronghorn. As you scan the prairie, the complexity of this habitat mosaic may not initially be apparent. Both parks lie at an ecological crossroads where the Great Plains stretch toward the Rocky Mountains. In this transitional zone, forests of ponderosa pine reach their eastern boundary, and American elm and bur oak are at their westernmost range. Western wood-pewees, Townsend's solitaires, and mountain bluebirds, birds of the coniferous forest, are common residents. Species found in eastern woodlands such as red-headed woodpeckers and eastern king-birds migrate this far west for the summer. Within the expansive midgrass prairie as well, eastern and western species meet in the diversity of forbs and grasses and pockets of summer-blooming wildflowers.

The Fast and Formidable

Unlike bison, which boldly saunter across the prairie, elk are reclusive and shy, yet no less majestic and powerful as the bison. In spring and summer, they seek secluded woodlands, emerging at dusk to forage until early morning. Never venturing far from the forest, elk favor those areas where forested slopes merge with open grasslands. Unlike buffalo, which graze unperturbed, rarely lifting their heads to look around, elk are attentive and watchful, looking up, looking around, always vigilantly checking their surroundings. Once the most widespread of all hoofed animals in North America, elk numbered in the millions. Today, Wind Cave supports a herd of around 300, Custer State Park close to 1,000.

These truly formidable creatures are considerably larger than their relative the deer. In the family of cervids, to which they both belong, only the moose is larger. Bulls stand five or six feet tall and weigh up to 1,000 pounds. Their antlers, bearing a dozen tines, six on each side, extend several feet into the air. The males lose their antlers in spring, then grow them again in summer, as fast as a half-inch per day, coated with a velvety, blood-rich tissue.

the race was finished, the massive weight of all those pounding feet caused the earth to heave and buckle. Flames erupted, ash spewed forth, and gnarled and twisted rocks bulged in the center. When the smoke cleared and the debris settled, the Black Hills as we now know them had been formed. Leading from the granite core to the hogback ridge circling the hills were low slopes alternating between deep forests and lush grasslands.

The dense, interlocking mosaic of shadow and sunlit space is ideally suited to all sorts of creatures: reptiles and amphibians, nearly 200 species of resident and migratory birds,

As fall approaches, it's time for the rut, when elk are easier to observe. Bulls rub their racks against trees to polish and sharpen the points and clash head-on. The victors breed with as many females as they can handle, calling to them with an eerie, high-pitched bugling, audible for long distances. Rival bulls strut before engaging in battle. They thrash small trees with their antlers, bugle long and loud, trot sideways to show off their size and muscles, and spray urine like a firehose. If the contest continues, they lock antlers with a loud *thwock!* Once a bull gains the uphill advantage, he uses his weight and momentum to drive his opponent down. The encounter sometimes proves fatal. Be sure to stop at the Wind Cave visitor center to see the pair of skeletons with locked horns – all that remains of two bulls that failed to disengage and perished of starvation.

To fully appreciate the pronghorn, also known as the American antelope, you need to see it in full flight. The fastest mammal in North America, pronghorns can speed across any kind of terrain. Their muscular haunches, tapered legs, and oversized lungs and heart enable them to cruise at 30 miles per hour, bound in 20-foot leaps, and accelerate up to 60 miles per hour in short bursts – an advantage in escaping predators. Their most ardent predator, the gray wolf, was common in the Hills in the 1800s but has not been seen since 1928.

Pronghorns are not difficult to spot in the field. Their coloring is distinctive – reddish tan along the back and haunches, creamy chest and belly, zebralike striations on the neck. When alarmed, their white rumps flash like signal flags. The horns of adult males are shaped like jagged sickles. Females also have horns, which are shorter and straighter. Pronghorns browse for forbs and dwarf shrubs in areas where the grass has already been thinned by grazing bison. Look for them along the 18-mile **Wildlife Loop Road**

Ferruginous hawks (above) often stand on the ground to wait for mice, squirrels, and gophers to scuttle out from their burrows.

Cathedral Spires (below) stand in the northwest corner of Custer State Park. Mountain goats, introduced in the 1920s, are found in this area.

Prairie Symbiosis

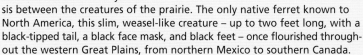

The black-footed ferret and the prairie dog – theirs is a relationship that dramatically illustrates the delicate symbiosis between the creatures of the prairie. The only native ferret known to North America, this slim, weasel-like creature – up to two feet long, with a black-tipped tail, a black face mask, and black feet – once flourished throughout the western Great Plains, from northern Mexico to southern Canada.

As the number of prairie dogs declined, so too did that of the black-footed ferret, a nocturnal prowler that feeds almost exclusively upon the rodents. The ferret was listed as an endangered species in 1967, but, by the mid-1970s, most wildlife biologists thought it was destined for extinction.

Black-footed ferrets (right) are smaller than prairie dogs, but consume the large rodents and live in their burrows.

Prairie dogs (below) are social animals that use 10 different calls to sound warnings, defend territory, or show aggression.

Then, in 1981, a remnant group of ferrets was discovered on a ranch in Wyoming. In the late 1980s, after a plague nearly wiped them out, the last 18 ferrets were trapped and removed to a captive breeding facility. Since 1991, state and federal government agencies, in conjunction with private organizations, have been actively reintroducing black-footed ferrets into the wild at six different sites in the West. Success of the program depends on a healthy population of prairie dogs, a species that continues to suffer losses to ranching and development. If prairie dog numbers remain stable, biologists hope to have 1,500 ferrets established in the wild by 2010.

in Custer State Park and the **Bison Flats** and **Red Valley** area of Wind Cave.

Mule deer and white-tailed deer abound in the Black Hills. Numerous trails wind through the trees and grasslands of both Wind Cave National Park and Custer State Park, and you can also take off cross-country to find a good vantage from which to observe deer and other wildlife. You are more likely to see mule deer in open areas rather than dense forests. They are instantly recognizable by their remarkably long ears and black-tipped tail. Should you startle a mule deer, most likely your yelps of surprise will mingle with the sound of its bounding feet as it lets out a pronounced *chuff!* and bounces out of sight as if on a pogo stick. Whitetails, smaller than mule deer, have white undertails, and excellent vision and

sharp hearing. In contrast to mule deer, they prefer woodland habitat with plenty of cover. Deer of both species tear off vegetation in ragged patches like rabbits, rather than biting it clean, and you may notice this sign as you explore both parks.

Two major species once indigenous to the area are gone from both parks: wolves and grizzly bears. Mountain lions, the sleekest, most efficient killing machines in nature, are rarely seen. If you spot one, count yourself lucky – and keep your distance. Audubon bighorn sheep once frequented the Black Hills but disappeared by the turn of the century. Rocky Mountain bighorns, a close relative, were introduced to Custer State Park in 1922, then again in 1964. The population, currently at about 150 animals, can be found along **French Creek**, a rugged watercourse that threads through the center of the park, and on the west side of the park near **Mount Coolidge**. These agile, high-wire acrobats are adept at disguising their appearance. Their rufous coloring and ability to remain perfectly still enable them to blend into the grass and rocks.

A Dog's Life

The black-tailed prairie dog is one of the foundation species of the Great Plains ecosystem. Intensely social, prairie dogs rely on one another for security and companionship. They live in complex tunnels with mounded entrances where they spend much time frolicking, eating grass, and watching for predators. Their vocalizations run the gamut from repetitive low-keyed chatter to piercing whistles of alarm.

Encountering these hyperactive, golden-brown rodents during their 1804–06 expedition, Lewis and Clark recorded their discovery, using the name *petit chiens,* or "small dogs," after an older term bestowed on the animals by French-Canadian trappers. Though the name persists, prairie dogs actually belong to the squirrel family. The species, once numbering in the millions, ranged over thousands of square miles of unspoiled grasslands. Over the past 100 years, conversion of grasslands into farmland, grazing pasture, and suburban development has reduced prairie-dog habitat to less than one percent of its original size.

No other creature in the grasslands system is as vital to the continued health of so many other prairie species. Coyotes, badgers, and bobcats feed on them, as do golden eagles and red-tailed hawks. Burrowing owls and prairie rattle-snakes share their subterranean lairs. Turkey vultures, black-billed magpies, and crows scavenge their dead carcasses. The creature most symbiotically connected to the prairie dog, the black-footed ferret, used to inhabit their burrows and keep their prolific numbers in check, but is now engaged in its own perilous struggle for survival.

Two of the best places to observe prairie dogs are at the southern end of the Wind Cave **visitor center access road** and four miles south of the **Blue Bell Area** in Custer State Park. Predators roam the sky and prowl the edges of these extensive prairie-dog towns spread out across the rolling hills. Deer mice scurry for cover. Western meadowlarks fill the air with flutelike jingles. The horned lark makes its nest in shallow ground cups, lined with grass and feathers.

A half-dozen pronghorns, in a ragged file, stepping with balletic grace, browse for broadleaf plants between clusters of squeaking prairie dogs. The antelope remain alert, taking in their surroundings with their wide-set eyes, liquid and protuberant. Nearby, a mighty bison rolls across the mounds and grassless patches to relieve the itchy sting of insect bites. The weight of these shaggy behemoths compacts the grass and complex root system, which prairie dogs then open up with their indefatigable digging. Spend a little time here, looking through binoculars, noting the colors and movements, listening to the wind and the chorus of grunts and squeals, and you will begin to understand how fragile and compelling all these links really are.

Pronghorns are unique to North America. Evidence suggests that they lived on the continent's plains and deserts a million years ago.

DETAILS

When to Go

Summer is warm and pleasant, with highs in the 90s and occasional afternoon thunderstorms. Spring and fall are brisk and breezy. Expect heavy snowfall in winter. Visitors may see sparring bison or hear bugling elk during the rut in summer and early fall.

How to Get There

The Rapid City Regional Airport is located about 30 miles from Custer State Park and 60 miles from Wind Cave National Park. Airport Express, 800-357-9998, can arrange shuttle service between the airport and both parks.

Getting Around

Car rentals are available at the airport. Bison, pronghorn, and prairie dogs are visible from the road, but it's always advisable to try at least a short hike on one of the well-marked trails. Custer State Park also has horseback and mountain bike trails.

Backcountry Travel

A free backcountry camping permit is required at Wind Cave and is available at the visitor center. Self-registration is required at Custer's two primitive campsites, located in the French Creek Natural Area.

Handicapped Access

Visitor centers and select campsites and lodgings are accessible.

INFORMATION

Custer State Park

HC 83, Box 70; Custer, SD 57730; tel: 605-255-4515.

Hot Springs Area Chamber of Commerce

801 South 6th Street; Hot Springs, SD 57747; tel: 800-325-6991 or 605-745-4140.

South Dakota Department of Tourism

711 Wells Avenue; Pierre, SD 57501; tel: 605-773-3301.

Wind Cave National Park

Hot Springs, SD 57747; tel: 605-745-4600.

CAMPING

Custer State Park has seven campgrounds and 325 campsites, some available on a first-come, first-served basis, others by reservation. To reserve a campsite, call 800-710-3267. Wind Cave National Park has one campground and 75 campsites, available on a first-come, first-served basis.

LODGING

PRICE GUIDE – double occupancy

$ = up to $49	$$ = $50-$99
$$$ = $100-$149	$$$$ = $150+

Angostura State Resort

HC 52, Box 125; Hot Springs, SD 57747; tel: 800-364-8831 or 605-745-6665.

These four modern cabins are situated on scenic Angostura Reservoir and offer a panoramic view of the Black Hills. Private cabins have two bedrooms, two full baths, a living room, and a full kitchen. A heated swimming pool, a furnished patio, and boat rentals are also available. Open April through October. $$$

Blue Bell Lodge and Resort

Custer State Park Resort Company; HC 83, Box 74; Custer, SD 57730; tel: 605-255-4531 or 800-710-2267.

Built in the early 1920s by a Bell Telephone executive – hence the name – the lodge is surrounded by ponderosa pines and situated along French Creek at the base of Mount Coolidge. The resort has 29 cabins, with fireplaces and open-beamed ceilings; some are more rustic than others. Restaurant dining, western chuck wagon cookouts, hayrides, horseback riding, and pack trips are also available. $$–$$$

Historic Franklin Hotel

700 Main Street; Deadwood, SD 57783; tel: 800-688-1876 or 605-578-3452.

Guests at this gracious hotel have included Teddy Roosevelt, Babe Ruth, Pearl Buck, and Kevin Costner. Opened in 1903, 11 years after its foundation was laid, the hotel recently underwent a five-year, $2 million renovation. Its four-story Greek Revival architecture is magnificently preserved, while the interior features an assortment of precious woods, ceramic mosaic tiles, fluted columns, and pressed tin. The hotel has 14 luxury suites, 23 historic rooms, 23 economy rooms, and 13 motor inn rooms. Amenities include a restaurant, casino, and sports bar. $–$$$$

Larive Lake Resort

Box 782; Evans Plunge, SD 57747; tel: 605-745-3993.

In addition to six modern cabins, this resort, 30 miles southeast of Custer State Park, offers 20 tent sites and hookups for recreational vehicles. Paddle boats and a game room are available.

State Game Lodge

Custer State Park Resort Company; HC 83, Box 74; Custer, SD 57730; tel: 605-255-4541 or 800-710-2267.

This lodge served as President Calvin Coolidge's "Summer White House." Built in 1920, the stone-and-timber building is set in a beautiful mountain valley surrounded by pine forest. Seven rooms retain their stately historic charm; 40 separate motel units afford basic comfort. The lodge also has cabins, some with kitchenettes. All guest rooms and cabins have private baths. A restaurant, a lounge, and safari Jeep rides are also offered. $$–$$$$

Sylvan Lake Resort

Custer State Park Resort Company; HC 83 Box 74; Custer, SD 57730; tel: 605-574-2561 or 800-710-2267.

This stone-and-timber lodge overlooks Sylvan Lake and Harney Peak. The resort has more than 60 rooms and 30 rustic cabins. All units have a private bath; some cabins have a fireplace and full kitchen. A restaurant and boat rentals are available. Cabins are open from May 1 to September 27, the lodge from May 1 to October 5. $$–$$$$.

TOURS & OUTFITTERS

A-Z Adventures

4068 Canyon Drive; Rapid City, SD 57702; tel: 605-343-7981.

Fully outfitted, custom-designed horseback riding, mountain biking, and hiking excursions in the Badlands.

Custer State Park Resort Company

HC 83, Box 74; Custer, SD 57730; tel: 800-658-3530 or 605-255-4772.

Hayrides, chuck wagon cookouts, horseback trail rides (day or overnight), safari Jeep rides, as well as rentals of boats and mountain bikes.

Gunsel Horse Adventures

P. O. Box 1575; Rapid City, SD 57709; tel: 605-343-7608.

Four-day pack trips in spectacular Badlands National Park, featuring wildlife watching, old homesteads, prehistoric fossils, and Native American lore. Meals and tents included.

Excursions

Badlands National Park

P.O. Box 6; Interior, SD 57750-0006; tel: 605-433-5361.

The prairie takes on a completely different aspect about 60 miles east of the Black Hills. This is the Badlands, an immensely rugged country of furrowed cliffs, gnarled spires, and deep, branching ravines torn from the plains by a half-million years of erosion. Haunting the dramatically desolate park are eagles, owls, vultures, coyotes, prairie rattlers, bison, pronghorn, and bighorn sheep. Wildlife watching is particularly good in the Sage Creek Wilderness Area.

Black Hills National Forest

RR2, Box 200; Custer, SD 57730-9501; tel: 605-673-2251.

Mount Rushmore's four august faces give a solemn air to the heart of this 1.2 million-acre forest, but the region is equally notable for an abundance and variety of wildlife. The Black Hills sit at an ecological crossroads where the Great Plains meet the mountains and eastern and western species overlap. Ponderosa pine at its easternmost limit grows near American elm at its westernmost. The western wood pewee and pinyon jay share the skies with the eastern bluebird and phoebe. And a wide range of mountain, forest, and prairie creatures – from mountain goats and bighorn sheep to elk and white-tailed deer – can be seen in a single visit.

Theodore Roosevelt National Park

P.O. Box 7; Medora, ND 58645; tel: 701-623-4466.

An early visitor once described these badlands as "Hell with the fires out." Later, in the late 1880s, Theodore Roosevelt tried his hand at ranching here, only to be driven out by severe weather. But there's a softer side, too. Songs of meadowlarks, sparrows, and towhees carry through prairies and woodlands. Golden eagles ride air currents overhead. Bald eagles migrate through in winter, and bison, wild horses, and white-tailed deer roam the grasslands. Predators such as bobcats, badgers, and rattlesnakes are present, too, though far less conspicuous.

Yellowstone National Park

Wyoming

CHAPTER 14

Dawn breaks clear and cold over the high country of Yellowstone. It begins with a brilliant red flush over the Absaroka Range, the far-off yipping of coyotes, the stirring of the first mountain breezes, a pair of patrolling ravens flying low and slow over the sagebrush. The mountain valley is wide and deep, the frosted floor hidden in shadow, the river reflecting the color of the sky.   ◆   Slowly, with a caution borne of experience, a band of bull elk step from a grove of golden aspen, black noses lifted as they test the air. There are five of them, ranging in size from a spindly four-pointer to a heavy-beamed monarch with seven ivory-tipped points on one side, eight on the other. Two of the younger bulls break from the others and gallop down the hill toward water. They stop and lock antlers, then trot off together, racks streaming over their shoulders.   ◆   A small group of people watch from the side of a road. Scanning the surrounding landscape with binoculars, someone spots a large black wolf loping across the sagebrush toward the bull elk, hidden by a ridge. The wolf moves rhythmically with visible strength, his head lifted and tongue hanging out. His tail is up. Then there are more wolves – three of them, colored like their father, all members of the Rose Creek pack – closing rapidly on the other flank. They accelerate, ears flattened, tails straight, legs a blur, a plume of dust behind them. The elk are still upwind, moving steadily down a worn bison trail, unaware of what is about to happen.

The world's first national park is the province of grizzly bears and gray wolves, moose and bighorn sheep.

Mammoth Hot Springs, near the park's northern entrance, is a prime spot to see bands of elk.

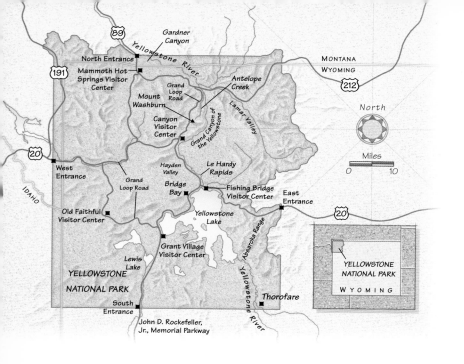

Reminders of an Earlier Time

This is the northern range of **Yellowstone National Park** in Wyoming, home to some of the densest populations of wild animals in the world. Formed in 1872 as the world's first national park, Yellowstone protects 3,472 square miles of high volcanic plateau at the headwaters of six major western rivers. The park is a vignette of the old western frontier, when elk, bison, and deer still roamed the mountains and valleys in natural abundance. Here in Yellowstone, visitors can still experience some of the sights that greeted the early fur trappers and explorers – a herd of bison spread across a grassy valley, a grizzly bear battling gray wolves over a moose carcass, bighorn sheep clambering along rocky ledges, a clear fast river filled from bank to bank with spawning cutthroat trout.

Add to this some of the most amazing geology known to science – geysers, geothermal pools, boiling rivers, colorful hot springs, petrified trees, a "Grand Canyon," legendary waterfalls, and the largest high mountain lake in North America – and you have a pristine region the likes of which cannot be found anywhere else on the planet.

Seasonal Hot Spots

Because of the size of the park and the diversity of fauna, you need to know in advance where and when to go for the best wildlife watching. The optimal season for large predators is spring, from late April through early June, when grizzlies and wolves hunt for newborn ungulates, such as moose and elk, or adults weakened by the long winter. Summer brings migratory waterfowl, songbirds, and raptors, as well as the spawning of Yellowstone cutthroat trout, visible in June and July at viewing locations near **Le Hardy Rapids** and around **Fishing Bridge**. This spawning run attracts its share of hungry grizzly bears, some of which have become adept at catching trout. Autumn is the best time for antlered animals – elk, deer, and moose – when their racks are fully formed and the rut, or mating season, is at its peak.

Wildlife is abundant just about everywhere in Yellowstone, but like any ecosystem, the park has its hot spots, where food, cover, and water regularly concentrate wildlife. The area around **Mammoth Hot Springs**, for example, is one of the finest places, year-round, to view bighorn sheep, especially in **Gardner Canyon** below the campground, as well as elk. The local elk herd often lounges in the grass outside the park headquarters building. Visitors are often shocked to find a bugling bull elk standing between them and the park post office, or a cow elk nursing her calf beside their picnic table, or two bull elk locking antlers a few steps from the park library.

Another superb area is **Lamar Valley**, 12 miles east of Mammoth Hot Springs on the park road. It is possible in one drive through Lamar Valley to see nearly all of the large mammals in the park, from grizzly bears to wolves, bison to antelope, elk to deer. The expansive sage and grass meadows, the lazy winding river, and the gentle aspen-covered ridges all make for a wildlife-viewing paradise. Unlike other areas in the Rocky Mountains, where animals are seen only at dusk or dawn, they can be observed in Lamar Valley at virtually any time of day.

Five miles south of Lamar Valley, the Grand Loop road climbs a series of switch-backs toward a range of mountains west of the Yellowstone River. Spruce and fir darken the slopes, and aspen groves dot the land-scape. Clear streams tumble down mountain flanks. Grass meadows, nourished by rains, spread luxuriously. The area around **Antelope Creek** is the finest location outside Alaska for observing grizzly bears. Every spring, dozens of bears are drawn to the creek, which drains the north side of Mount Washburn, to hunt elk calves and to court and breed. Visitors can observe the complete span of the bears' daily activities: sleeping, nursing cubs, digging for plant roots, defending kill sites, grazing like cows on the pro-tein-rich spring grass, and mating (if the

Winter in Yellowstone

One of the first wildlife lovers to visit the park in winter was President Theodore Roosevelt. Arriving in early April 1903, he found much of Yellowstone still covered with snow and had to abandon his horse and climb aboard a horse-drawn sled. Cross-country skis, popular even then, enabled the Rough Rider to reach the higher regions of the park, where he marveled at the bison and elk wintering in the geyser basins.

Bison (left) plow their massive heads through the snow to reach buried plants. They are seen in groups near geyser basins and streams free of ice.

Bighorn sheep (below, left) blend into their surroundings during much of the year, but in winter they stand out against the snow.

Mountain lions (below, right), river otters, coyotes, and other animals active in winter all leave identifiable tracks.

Visitors today can see much the same sights from December through March, whether entering the park from West Yellowstone, the west entrance, or Gardiner, known as Gateway to the Northern Range. Many roads in the park interior are open for snowmobiles and snow cats, or "snow-coaches." A popular route runs

from Gardiner south to Old Faithful, in Geyser Basin. The highway between Gardiner and Cooke City, on the eastern side of the park, is maintained for cars year-round. Cross-country skiers and snowshoers will find a wealth of trails, each easily identified by orange metal markers located well above snow level.

Winter wildlife in the park is abundant. Expect to see over-wintering waterfowl – herons, geese, ducks, swans – along rivers warmed by geothermal springs, such as the Firehole. Bison and elk often gather at the geyser basins, and ravens, coyotes, and wolves are not far away, waiting for their next meal. Gardner Canyon, north of Mammoth Hot Springs, just outside the park boundary, is an excellent place to view wintering herds of bison, bighorn sheep, mule deer, antelope, and elk.

In the winter months, particularly energetic auroral displays are visible at the latitude of Yellowstone. Park visitors can experience all the glory of Alaska and the Yukon as the ghostlike northern lights, or aurora borealis, flicker over the snow-covered wilderness.

females seem huge, the size of the males will astonish). It is not unusual to find shed grizzly-bear fur on park road signs, as grizzlies rub off their heavy winter coats. Lucky visitors can touch the long silver-tipped strands as soft as silk and smell the musky ursine odor that lingers long on the fingers.

Venturing into the High Country

Mount Washburn, four miles south of Antelope Creek, is a heavy, solid mountain, the remnant of a million-year-old volcano. It is a shaggy old massif with Douglas fir, whitebark pine, and fireweed on its gentle spreading slopes. A trail leading up the northwest ridge takes hikers away from the busy road and into the quiet Yellowstone backcountry. Experienced hikers make the climb in late morning, when the resident grizzlies have stopped feeding and retired to the woods for the day. Along the trail there is powerful evidence of

the 1988 fire that swept through Yellowstone, burning up to a third of the park's forests. A wall of fire swept over the ridge, killing most of the whitebark pines. These trees the grizzlies sorely miss, for the pine nuts are an essential part of their diet in autumn. On Mount Washburn, though, as elsewhere in the park, the fire has had a beneficial effect, opening forests and letting sunlight reach the green vegetation on which the animal kingdom depends.

Along the way there are always scattered bands of elk and mule deer – shy at first and eager to run off if too much interest is shown. And at the top (10,234 feet) there is a prospect to take the breath away: the far-off blue and gray peaks of the Absaroka Range, the Grand Canyon of the Yellowstone, the geyser basins with their steam vapor clouds, a bit of cobalt-hued Yellowstone Lake, and everywhere mountains and valleys without end. Naturalist John Muir, who made the ascent in 1895, spending the night on the

Return of the Wolf

In the spring of 1995, following years of careful planning, 31 gray wolves were imported from Canada and released in Yellowstone National Park. Since then, the wolves have spread through the forests and meadows of the park and formed 10 breeding packs, each with its own home territory. At last count there were more than 100 wolves in the park, including adults, yearlings, and pups. Four packs – the Chief Joseph, Leopold, Rose Creek, and Druid Peak – inhabit the northern range of the park and, because of the wide-open spaces and year-round accessibility by road, are among those wolves most easily seen. This is especially the case in the late winter and early spring when deep snows drive the elk, deer, antelope, moose, and bighorn sheep from the high plateau into Lamar Valley and Gardner Canyon. Although off-road hiking, snow-shoeing, and cross-country skiing are restricted at this time to avoid stressing the animals in the valley and canyon, the resident gray wolves can be observed and photographed from the road.

Gray wolves (left and below) are social animals that live in packs, the males and females mating for life. Most wolves are indeed gray, but black is one of the species' color phases.

Bull elk (opposite) when mature are about five feet high. The spread of their antlers can be wider than the animals are tall.

summit to watch the stars and then observe the sunrise, waxed poetic about the experience: "Where else may the mind find more stimulating, quickening pasturage? A thousand Yellowstone wonders are calling from the summit of Mount Washburn!"

Hikers who linger on the summit until late afternoon may observe bighorn sheep, the sturdy, sharp-eyed, indefatigable cliff dwellers that make a home of this remote windswept place. The Rocky Mountain bighorns inspire a certain awe. There is something enchanting about their massive horns, something about the circular sweep that attracts the human longing for the symmetrical in nature. There is something, too, about where they live – the lichen-covered rocks and terraced ledges, the alpine meadows and timberline forests, the places that have been set so far above the rest of the world. It is a hard life, with long winters and summer lightning and wolves that love mutton and lamb more than any other food.

Where Ospreys and Eagles Soar

Fifty miles to the south of Mount Washburn, among the secret springs and melting snow-fields of the plateau, the **Yellowstone River** begins. After leaving the great lake, the clear deep river runs level for nearly 20 miles through a sun-filled valley fringed by lodgepole pine forests. There are herds of bison and elk along its course, as well as blue herons and sandhill cranes, Canada geese and wood ducks, white pelicans and trumpeter swans. River otters dip and swim, their slender bodies smooth and muscular and perfectly shaped for the water; industrious beavers ply the shores, ever on the lookout for trees to fell.

Then, with little warning, the river rushes through pools and past boulders with gathering strength, breaks over a thick slab of bedrock, and goes thundering off into the mist-filled gulf of the **Grand Canyon of the Yellowstone**. The first falls is 109 feet high. The second, a third of a mile downstream, is 308 feet. The whole of the canyon, 20 miles long and a thousand feet deep, is cut through bright volcanic ash and rhyolite lava of nearly every color imaginable: sun-bleached white, rosy pink, yellow ocher, rusted sienna, Mars orange, faded green, and blackish blue.

The canyon, 20 minutes south of Mount Washburn on the park road, is the summer home of the osprey, also known as fish

hawk, which nests in the rugged pinnacles near the rim and dives hundreds of feet to catch cutthroat trout in canyon pools. Ospreys live in mated pairs and return each year to the same carefully built nests, about a half dozen of which are visible in the upper part of the canyon. The entire population leaves at about the same time in late September or October, usually on the leading edge of a storm, and winters along the coasts of Mexico, Central America, and Venezuela. Golden and bald eagles inhabit the canyon as well, and, like the osprey, are often seen carrying fish aloft in their talons.

Shaggy Giants

Hayden Valley, three miles south of the Grand Canyon, has for most Yellowstone aficionados a magic ring. Here is the best place in the park to observe bison. On a foggy August morning, you can walk for miles in solitude across the grassy meadows, now in the broken sunlight of a hilltop, now

in the London pea soup of a sulphur-scented creek bottom, and never be far from the great beasts. You will hear them, smell them, and see them looming out of the fog. Their sign is everywhere, for Hayden Valley is an enormous pasturage where the herds come to graze throughout the summer. The bison are peaceful animals, except during the rut, and allow people to observe them, regarding visitors with calm umber eyes. The bison have little to fear from humans. They are the largest land mammal in the Western Hemisphere, weighing up to 2,000 pounds.

Every summer, a whirlwind of activity surrounds **Yellowstone Lake**, just south of Hayden Valley, as boaters, wildlife observers, and anglers converge on the largest high-altitude lake on the continent – about 20 miles long and 15 miles wide, with a coastline of over 100 miles. It is full of trout, both cutthroat and lake, and supports a multitude of birds, including swans, pelicans, geese, ducks, cranes, herons, curlews, and plovers.

You'll often see moose, elk, deer, and bear along the inlets. The extreme southeast and southern arms are restricted to hand-propelled craft (kayaks and canoes), and a marina and boat ramps can be found at Bridge Bay and Grant Village. Charter boat tours to **Stevenson Island** provide a chance to view nesting bald eagles, as well as gulls and ducks.

Wild Heart

If Yellowstone has a wild heart, it is the **Thorofare Country**, an immense roadless tract south of Yellowstone Lake, best reached from trailheads near Lewis Lake. Thorofare, at the head of the Yellowstone River, is horseman's territory, with blazed trails running 20 or 30 miles back into the plateau and streams and rivers impassable to foot travelers until early August. The biggest of the Shiras moose live here, as do red-eyed loons that call out eerily over the lakes as they do in the far north of the Yukon, and reclusive lynx and wolverine. This is also the home of fabled grizzlies known only by plaster casts made of their colossal tracks, as well as two of the park's largest and least-known wolf packs. Here wildlife lovers pass as through a wormhole and enter an older universe – the wild place from which we came, the wild place without which we cannot live.

In the end, Yellowstone National Park, like so many of the world's finest parks, is a place too large to be folded into a map or compressed into an essay. It is a special province of nature where geysers erupt with steaming water even in coldest January and snows fall on the highest peaks even in July, where dusty bison gather to drink at great rivers and gray wolves watch them with eyes like polished topaz, where bull moose lock antlers in cottonwood groves and bull elk bugle among golden aspen trees, where ospreys tuck their wings and dive for rising trout and white-hooded eagles soar through peaceful summer clouds on wings nine feet wide. It is a place to carry deep inside and hold close whenever you feel yourself at turmoil, and instantly be at peace.

Please Don't Tempt the Bears

Visitors to Yellowstone are guests in the home of the permanent residents: bears. Grizzlies prefer to avoid human contact, and observing some commonsense rules will ensure your safety in bear country.

Avoid surprising bears. Make your presence known by shouting, singing, or clapping, and hike in groups of three or more whenever possible. Keep the wind to your back so bears smell you.

Give bears their space. If you see fresh bear prints or droppings or evidence of bears feeding, choose another route.

Never feed bears. Always pack out all garbage.

Do not run if confronted. A bear may bluff its way out of a threatening situation by charging. When the bear stops, back away slowly at a diagonal.

Play dead if attacked. Roll into a fetal position with your hands behind your neck. Stay motionless to show the bear that you are no longer a threat.

Bison (opposite) cross the Firehole River, whose waters eventually flow into the geyser basins near Old Faithful.

Grizzly bears (right), as well as black bears, generally avoid people, but can be unpredictable and aggressive.

Bear tracks (above, right) and scat should be regarded as warning signs; you may be intruding on the bruin's territory.

TRAVEL TIPS

DETAILS

When to Go

The park is most inviting, and most crowded, in summer. Weather in July and August is warm and pleasant, with highs around 90°F and cool nights. Winter is harsh, with extremes well below 0°F. Most park roads are closed by snow from November to April.

How to Get There

Commercial airlines serve Cody and Jackson, Wyoming; Bozeman and Billings, Montana; and Idaho Falls, Idaho. Car rentals are available at the airports.

Getting Around

Bus tours are offered by AmFac Parks and Resorts, 307-344-7311, and National Park Tours-Grayline, 307-733-4325.

Backcountry Travel

A backcountry use permit is required for all overnight backpacking trips and can be obtained at most ranger stations and visitor centers no more than 48 hours in advance. For a campsite reservation form, call the Backcountry Office at 307-344-2160.

Handicapped Access

Visitor centers and some campsites. A list of accessible sites may be obtained from the Handicapped Access Coordinator, Box 168, Yellowstone National Park, WY 82190.

INFORMATION

Yellowstone National Park

P.O. Box 168; Yellowstone National Park, WY 82190-0168; tel: 307-344-7381.

Chambers of Commerce

West Yellowstone, 406-646-7701
East Yellowstone/Wapiti Valley, 307-587-9595
Jackson, 307-733-3316
Cody, 307-587-2297.

CAMPING

There are 12 campgrounds in the park; seven are open in summer and fall. Only one, Mammoth, is open year-round. Reservations are available for five facilities; call AmFac Parks and Resorts, 307-344-7311.

LODGING

PRICE GUIDE – double occupancy

$ = up to $49	$$ = $50-$99
$$$ = $100-$149	$$$$ = $150+

Canyon Lodge

P.O. Box 165; Yellowstone National Park, WY 82910; tel: 307-344-7311.

This modest wood-frame lodge offers attractive rooms with modern furnishings. The main building is flanked by 580 cabins, each containing four or more units with private baths. The complex is about half a mile from the Grand Canyon of the Yellowstone. Amenities include a restaurant, a cafeteria, a snack shop, horseback riding, and guided tours. Summer only. $$

Elephant Head Lodge

1170 Yellowstone Highway; Wapiti, WY 82450; tel: 307-587-3980.

Built in 1910 by Buffalo Bill Cody's niece, these 10 log cabins are set in Shoshone National Forest about 11 miles from the park's eastern entrance. Restaurant dining, horseback riding, and cookouts are available. Open May to October. $$

Lake Yellowstone Hotel and Cabins

P.O. Box 165; Yellowstone National Park, WY 82910; tel: 307-344-7311.

Set on Yellowstone Lake, this splendid four-story hotel, the park's oldest building, was erected in 1891 and recently renovated. Rooms are modern and offer lake or mountain views. One hundred cabins provide basic accommodations, each with two double beds, some with private bath. A lakeside dining room, a deli, a marina, and guided tours are available. Summer only. $$–$$$

Mammoth Hot Springs Hotel and Cabins

P.O. Box 165; Yellowstone National Park, WY 82190; tel: 307-344-7311.

The hotel offers 96 motel-style rooms just a short walk from Mammoth Hot Springs. Lodging includes 126 rustic cabins with shared baths. A restaurant and horseback riding are available. Open May to October and December to March. $$–$$$

Old Faithful Inn

P.O. Box 165; Yellowstone National Park, WY 82190; tel: 307-344-7311.

Completed in 1904, this 327-room lodge is the world's largest log building, with simple but comfortable accommodations and a monumental three-story fireplace in the lobby. The inn was designated a national historical landmark in 1987. Horseback riding and guided tours are available. Summer only. $$$$

Paradise Gateway Bed-and-Breakfast

P.O. Box 84; Emigrant, MT 59027; tel: 800-541-4113 or 406-433-4063.

This large cedar-shake house is set on 70 acres in the Absaroka Mountains, steps from the Yellowstone River and minutes from the national park. Four attractive guest rooms in the main house are decorated in French and English country style, each with private bath. Two separate log cabins have two bedrooms, full kitchen, laundry room, large living room, fireplace, and bathroom. Horseback riding, dog sledding, and guided tours can be arranged. $$–$$$

Yellowstone Suites

506 4th Street; Gardiner, MT 59030; tel: 800-948-7937.

About half a mile from the park's northern entrance, this three-story stone house offers three guest rooms and a suite, with Victorian decor, a shady garden balcony, and bay windows. The guest rooms have a shared bath; the suite has a private bath and kitchenette. Open year-round. $$

MUSEUMS

Museum of the Rockies

600 West Kagy Boulevard; Bozeman, MT 59717; tel: 406-994-3466.

An exploration of the greater Yellowstone ecosystem, including a world-class collection of dinosaur fossils.

National Bighorn Sheep Interpretive Center

907 West Ramshorn; Dubois, WY 82513; tel: 888-209-2795 or 307-455-3429.

Natural history of bighorn sheep, with emphasis on the life cycle and social behavior.

National Wildlife Art Museum

P.O. Box 6825; Jackson, WY 83002; tel: 307-455-3429.

Some of the nation's finest wildlife art in a gallery overlooking the National Elk Refuge.

TOURS & OUTFITTERS

Dozens of guides and outfitters work in the Yellowstone region. A list of outfitters licensed to operate in the park may be obtained from the Yellowstone Visitor Service, 307-344-2107.

Excursions

Glacier National Park

West Glacier, MT 59936; tel: 406-888-7800.

More than 200 grizzly bears range across the glacier-carved mountains and valleys of this glorious Montana park, which John Muir called some of the "best care-killing country on the continent." It's not unusual to spot shaggy mountain goats, bighorn sheep, moose, mule deer, and bald eagles, although elusive gray wolves, black bears, and mountain lions are more difficult to find.

Grand Teton National Park

P.O. Drawer 170; Moose, WY 83012; tel: 307-739-3300.

The Snake River meanders along the base of the stunning Teton range at this 485-square-mile park, providing habitat for moose, beavers, river otters, trumpeter swans, eagles, sandhill cranes, Canada geese, and other water-loving animals. Other species include elk, mule deer, and pronghorn antelope. There's excellent rafting on the Snake River, paddling in the lakes, and backcountry hiking.

National Elk Refuge

P.O. Box 510 ; Jackson, WY 83001; tel: 307-733-9212.

Each winter, up to 10,000 elk migrate to this 24,700-acre reserve, which adjoins the southern border of Grand Teton National Park in Jackson Hole. Visitors can watch the animals from the road or ride among the herd in a horse-drawn sleigh.

Rocky Mountain National Park

Estes Park, CO 80517; tel: 970-586-1333.

More than 78 peaks soar above 12,000 feet in this Colorado park, where placid lakes and fragile alpine tundra rival the dizzying heights. Mule deer, elk, moose, and bighorn sheep are a common sight, and the birding is quite good. Trail Ridge Road, the park's main thoroughfare, is one of the most scenic drives in the country.

Coronado National Forest
Arizona

CHAPTER **15**

t's late on a summer afternoon in the **Santa Catalina Mountains**, and a pair of hardy souls follow a narrow, rocky trail into steep-sided **Pima Canyon**. They set up a quiet observation post in the shade of a saguaro cactus, overlooking a small spring that sustains nearly every animal in the canyon. ◆ As the sun nears the western horizon and shadows lengthen, there is a *whoosh* of wings. A small flock of white-winged doves settles around the spring, cooing softly. Next come ravens and a scattering of summer tanagers and blue-gray gnatcatchers, then several grunting javelina, their disklike snouts held close to the ground. ◆ From the higher slopes, a band of desert mule deer descends the maze of pinnacles and talus at a deliberate pace. They are the color of weathered granite, their bodies smooth and hard. Pausing briefly to **Javelina, coatimundi, and many** nibble at brittlebrush or gaze at the depths **species of hummingbird** below, they move easily over what is **thrive where the Rocky** to them a familiar landscape. An old doe **Mountains and Sierra Madre** leads the way. Behind her follows a **meet the southern desert.** mixed group of does and immature bucks. At the spring, they gather and drink for a long time. At last the old monarch comes down, an immense buck with heavy antlers. He moves cautiously. The others glance over as he arrives, and then all lower their heads to the clear warm water. ◆ What makes this rich diversity of fauna and flora possible is the intermingling of three great geographic provinces: the Rocky Mountains, the Sierra Madre, and the low southern desert. Here, in **Coronado National Forest**, is found every climatic zone from subtropical desert to Canadian forest. In the course of one day, a hiker can climb from a landscape where cactus and creosote

The western patch-nosed snake speeds along the desert floor in pursuit of lizards, other snakes, and sometimes small mammals. It is often active during the day.

bush predominate to one where spruce and fir prevail, from a hot brushy oak canyon where several species of hummingbird nest to a windswept ridge where black bears leave their tracks in the snow. Beginning at the city limits of **Tucson**, Arizona, the national forest spreads in 12 sections across the southeastern part of the state, totaling almost two million acres. Each section protects a "sky island," or isolated mountain range, that rises from the surrounding sea of low desert.

Sky Islands

Unlike other parts of the American West such as the Rockies, where the outdoor season is restricted during the winter, Coronado National Forest offers year-round opportunities for wildlife viewing. Many birds that spend the summers on top of the sky islands are found during the winter months foraging along lowland streams. Species include yellow-eyed juncos, ruby-crowned kinglets, Bell's vireos, yellow-rumped warblers, and clay-colored and grasshopper sparrows.

Raptors are plentiful throughout the area at this time, especially red-tailed hawks and golden and bald eagles. Mule deer and Coues deer, a diminutive desert white-tailed deer, mate through December and January, a later rut than occurs farther north. Even on winter days, temperatures may reach the low 80s, making it possible to take a comfortable hike in the backcountry while the rest of the region is snowed in by blizzards.

Spring brings fabulous desert wildflowers, as well as the breeding season for many birds and animals: Gila woodpeckers, vermilion flycatchers, painted redstarts, Gambel's quail, coyotes, and gray foxes. Owls and hawks are abundant, as are migratory birds. Summer is the time for snakes and lizards. After a good rainstorm, reptiles can be found on trails and back roads. It is also the season for cactus flowers, which attract all manner of butterflies, moths, and birds. Good places for wildlife observation during the hot months are water holes, springs, and small creeks. Brief monsoons are common across the forest through July and August, when many people head into the high country.

Autumn is a favored season in the sky islands as the foliage turns and the deer start to lose the velvet on their horns. Wild grapes

ripen and black walnuts and acorns begin to fall, attracting skunks, bears, and other animals. Sandhill cranes arrive in mid-autumn and settle in for the winter. The return of winter brings the "vertical migration" as deer and other fauna descend from the mountaintops to the canyon bottoms, where they are concentrated for easier viewing.

The Wild Chiricahuas

Forty miles south and east of **Willcox**, Arizona, a sun-baked truck stop on Interstate 10, the **Chiricahua Mountains** rise from a low country of playas (dry lakes), washes (dry streambeds), and creosote flats. The Chiricahuas are a wild mountain range – topped by an 18,000-acre wilderness area – and a living remnant of the old Southwest. The entire range is part of Coronado National Forest.

Hiking the forest trails, such as the famous **Cave Creek Trail** near **Portal**, at dawn or dusk, visitors may see Coues deer, coyotes, hooded skunks, the indigenous piglike animals called javelina (also known as collared peccary), and the grayish brown coatimundi, a long-bodied, long-tailed relative of the raccoon. Small numbers of elegant trogon, year-round residents of Mexico, move north to nest here in summer and feed on the mast in the oak woodlands. Cactus wrens roost in the

Elf owls (opposite, above), the smallest owls found in North America, nest in the cavities of the saguaro cactus.

Short-horned lizards (above) and their reptilian relatives bask in the early-morning sun, then often retreat by midday.

Saguaro, ocotillo, cholla, and other cacti (left) sustain desert animals. Bats and hummingbirds harvest their flower nectar, and birds nest in their crooks and hollows.

The **Costa's hummingbird** (right) can be distinguished from other southwestern species by its purple head and throat.

Gambel's quail (below), permanent residents of southwestern deserts, are often heard before they are seen. Their constant clucking helps members of a covey maintain contact.

Javelinas (opposite), native piglike animals, bear young that within one day are able to scamper after the adults.

Feathered Jewels

Sixteen species of hummingbird can be seen in southern Arizona, more than in any other state, primarily because of the mild subtropical winters and the numerous nectar-rich flowers that bloom in spring and summer, such as coral bean, desert honeysuckle, scarlet sage, and penstemon species. Coronado National Forest is a permanent or temporary home to many species: Broad-billed, blue-throated, magnificent, rufous, broad-tailed, Costa's, Anna's, and black-chinned are commonly seen in Madera and Sabino canyons and the Upper San Pedro River Basin. White-eared, Lucifer, and plain-capped starthroat are somewhat rarer. The best times to observe hummingbirds are April, when a major pulse of northern migration passes through the state, and August into September, when there is a second bloom after the summer monsoons.

crooks of cholla cactus, and Lewis' woodpeckers frequent the oaks, especially along streams. Hikers reach the high country by following any of the major trails from Portal. Up higher in the Chiricahuas, the coniferous forests and flower meadows offer superb butterfly and hummingbird habitat during the summer months when the lower desert is dry and desiccated.

Flocks of thick-billed parrots, another resident of Mexico, were once commonly seen and heard screeching throughout the mountain forests and canyons of Arizona. Now extremely rare, parrots have been reintroduced in small numbers into the Chiricahuas, where a fortunate watcher might spot them. Although Mexican wolves and southwestern grizzlies no longer roam the Chiricahuas, jaguars still haunt the ridges and canyons in the area. A jaguar was killed in the **Bowie Range** of the Chiricahuas in 1987. More recently, in 1996, one was treed in the **Peloncillo Mountains** to the east and photographed by a local rancher, who has since taken the lead in saving this endangered species. Other sightings have been confirmed in the **Baboquivari Mountains** west of **Nogales**.

Thirty miles to the west of the Chiricahuas, across **Sulphur Springs Valley**, rises a much lower and more rocky range: the **Dragoon Mountains**. It was here that the Chiricahua Apache made their last stand against the early miners, settlers, and army troops. **Cochise Stronghold Canyon**, where the Apache chieftain and his warriors lived, lies in the beautiful wooded heart of the Dragoons. Hikers enter this canyon of huge granite domes and sheer cliffs on a self-guided nature trail from the campground at the end of Forest Service Road 84. The trail provides good opportunities for viewing birds, including acorn woodpeckers, ash-throated flycatchers, Cassin's and western kingbirds, common poorwills, Lucy's warblers, and western bluebirds. Coues deer, as well as coatimundi and javelina, appear at dawn and dusk. Hikers who follow the trail through the canyon to **Cochise Peak** are treated to a view that encompasses the historic town of Tombstone and the distant blue Sierra Madre of northern Mexico.

Cross Section of Habitats

Without a doubt, the best driving tour in Coronado National Forest is **Mount Lemmon Road**, which leads from northeast Tucson to the summit of **Mount Lemmon** in the Santa Catalina Mountains and offers plenty of places to stop and explore. The road passes through every life zone and animal habitat area in this part of the Southwest. For the first several miles, as the road climbs the steep ridges above **Soldier Canyon**, Sonoran desert vegetation – ocotillo, saguaro cactus, brittlebrush – predominates. This is the home of the desert tortoise and the western diamond-backed rattlesnake, the roadrunner, and the desert cottontail. Beyond **Molino Basin**, the road passes through the oak and juniper woodlands, which support gray-breasted jays (another species commonly found in Mexico), scrub jays, coatimundi, javelina, and gray fox. The **Arizona Trail** from **Molino Basin Campground** takes hikers into this area, where mountain lion and even black bear are occasionally seen.

Higher up, the oak woodlands merge into the ponderosa pine forest, one of the most productive ecosystems in the area, where endemic species like Coues deer occur in greater numbers. At **General Hitchcock Campground**, hikers can take the **Green Mountain Trail** into the canyonlands. The last zone, near the summit of 9,157-foot Mount Lemmon, is a mixed conifer forest. Here, amid the quaking aspen, Douglas fir, and ponderosa pine, are thick ferns and golden columbine, sunny

meadows and darkly shaded mountain glades with Abert's squirrels and, occasionally, a black bear. At road's end, trails lead to two natural areas, **Butterfly Peak** and **Santa Catalina**, in the upper reaches of Mount Lemmon. From this vantage, you can spot a peregrine falcon flying to and from its rocky aerie, or a northern goshawk startling songbirds feeding in the treetops.

In Coronado National Forest, you can experience many things throughout the seasons. You can hear a cactus wren singing at dawn from the top of a saguaro and a hermit thrush performing its evening matins from the branch of a Douglas fir. You can peer at a coatimundi drinking from a *tinajas* at the base of the Pinaleno Mountains and a black bear eating raspberries on the back side of Chiricahua Peak. You can see a fast-moving winter blizzard powder the mountains and shimmering summer heat waves lift the Santa Catalina range from the bonds of the earth – and toward the sky.

TRAVEL TIPS

DETAILS

When to Go

Head for the high country in summer, when temperatures at lower elevations often exceed 100°F. Be prepared for thunderstorms. The desert is much more inviting from October to May, when high temperatures are typically in the 60s and 70s. Night-time temperatures, however, can fall below freezing.

How to Get There

Commercial airlines serve Tucson International and Phoenix Sky Harbor International.

Getting Around

A car is essential. A vehicle with four-wheel-drive may be necessary on unpaved backcountry roads. Rentals are available at the airports.

Backcountry Travel

Permits are not required for backcountry hiking or camping.

Handicapped Access

Sabino Canyon Visitor Center and 18 recreation sites, including some campgrounds, are wheelchair accessible.

INFORMATION

Coronado National Forest

Federal Building, 300 West Congress; Tucson, AZ 85701; tel: 520-670-4552.

Greater Phoenix Convention and Visitor Bureau

400 East Van Buren, #600; Phoenix, AZ 85004; tel: 877-225-5749 or 602-254-6500.

Metropolitan Tucson Convention and Visitors Bureau

130 South Scott Avenue; Tucson, AZ 85701; tel: 800-638-8350 or 520-624-1817.

CAMPING

The forest has 35 campgrounds and 603 campsites. Most sites are available on a first-come, first-served basis. To make a reservation, call 800-280-2267.

LODGING

PRICE GUIDE – double occupancy

$ = up to $49 $$ = $50-$99

$$$ = $100-$149 $$$$ = $150+

Arizona Inn

2200 East Elm Street; Tucson, AZ 85719; tel: 520-325-1541 or 800-933-1093.

This sprawling adobe hotel was built in 1930 and is listed on the National Register of Historic Places. It is set on 14 acres of immaculate lawns and gardens and has 80 handsome rooms with private baths. Amenities include a restaurant, wood-paneled library, lounge, evening pianist, heated outdoor pool, and clay tennis courts, $$-$$$$

Bisbee Inn

45 OK Street, P.O. Box 1855; Bisbee, AZ 85603; tel: 888-432-5131 or 520-432-5131.

This scrupulously restored miner's hotel, built in 1917, is within easy walking distance of downtown Bisbee's historic district. Sixteen of the inn's 20 guest rooms have private baths. Three suites have private baths and kitchens. The inn's Victorian decor is supported by original antiques and hand-sewn quilts. Full breakfast included. $$-$$$.

Circle Z Ranch

P.O. Box 194; Patagonia, AZ 85624; tel: 520-287-2091.

Guests stay in seven adobe cot-

tages with 27 rooms, about 60 miles south of Tucson at the foot of the Patagonia and Santa Rita Mountains. The main activities are horseback riding, pack trips, and birding at the adjoining Nature Conservancy preserve. There is a three-day minimum stay. $$

El Presidio Bed-and-Breakfast Inn

297 North Main Street; Tucson, AZ 85701; tel: 800-349-6151 or 520-623-6151.

First-class lodging in downtown Tucson's El Presidio Historic District. Built in 1886, this Victorian adobe has three elegant suites with fine antiques, original art, private baths, and kitchenettes. Suites open to a brick-and-cobblestone garden courtyard, with rich floral displays and a three-tiered fountain. $$-$$$.

Portal Peak Lodge

P.O. Box 364; Portal, AZ 85632; tel: 520-558-2223.

Just south of Chiricahua National Monument, the lodge's 16 rooms are attractively furnished in Southwestern style. Rooms open to a long deck, with magnificent views of the Sky Islands. A restaurant, grocery store, and gift shop are on the premises. $$.

Tombstone Boarding House Bed-and-Breakfast

108 North 4th Street, Box 906; Tombstone, AZ 85638; tel: 520-457-3716.

Built in the 1880s and remodeled in the 1930s, these adjacent adobes occupy a quiet spot one block from town. Seven guest rooms, each with private bath, are furnished with antiques and collectibles associated with Tombstone's Old West history. A small, two-room miner's cabin has a private bath. $$

TOURS & OUTFITTERS

High Sonoran Adventures

10628 North 97th Street; Scottsdale, AZ 85260; tel: 602-614-3331.

Customized wilderness adventures – hiking, mountain biking, and whitewater rafting – in Coronado, Coconino, Kaibab, and Tonto National Forests and Grand Canyon National Park. Professional guides and equipment are supplied. Groups or individuals are welcome.

Sabino Canyon Tours

5900 North Sabino Canyon Road; Tucson, AZ 85750; tel: 520-749-2327 or 520-749-2861.

A narrated open-air bus tour of the Catalina Mountains, with opportunities for birding, swimming, hiking, and picnicking.

Tucson Audubon Society

300 East University Boulevard #120; Tucson, AZ 85705; tel: 520-578-1330.

Society volunteers lead free birding field trips to numerous locations in southeastern Arizona. Call for tour schedules.

PARKS

Chiricahua National Monument

Dos Cabezas Route, Box 6500; Willcox, AZ 85643; tel: 520-824-3560.

A fascinating mix of American and Mexican wildlife inhabit the park's 17 square miles, set amid the bizarre, eroded rock formations of the Chiricahua Mountains.

MUSEUMS

Arizona Sonora Desert Museum

2021 North Kinney Road; Tucson, AZ 85743; tel: 520-883-1380.

A "living" museum specializing in the shy, nocturnal creatures and unusual plants of the Sonoran Desert.

Excursions

Cabeza Prieta Wildlife Refuge

1611 North 2nd Avenue; Ajo, AZ 85321; tel: 520-387-6483.

Few places in the country are as wild and desolate as this 1,000-square-mile refuge. Situated in the southwest corner of Arizona, Cabeza Prieta was formed in 1939 after the Boy Scouts led an effort to save the desert bighorn sheep. Since then, the refuge has also served as a safe haven for the rare Sonoran pronghorn antelope and a variety of unusual plants, among them the scaly-barked elephant tree and senita cactus. Other notable creatures include the kangaroo rat, ground squirrel, and horned toad.

Organ Pipe Cactus National Monument

Route 1, Box 100; Ajo, AZ 85321; tel: 520-387-6849.

This 516-square-mile park on the U.S.-Mexico border protects a wide variety of flora and fauna but is best known for the many-armed organ pipe cactus. The park extends over low mountain ranges, vast arid plains, and sere salt flats. Twenty-eight types of cacti are found here along with bighorn sheep, coyotes, kangaroo rats, desert tortoises, snakes, and hawks. The area was designated an International Biosphere Reserve in 1976.

Saguaro National Park

3693 South Old Spanish Trail; Tucson, AZ 85730; 520-733-5153 or 520-733-5158.

A stone's throw from Tucson, the park's two districts encompass the Tucson and Rincon Mountains, east and west of the city, preserving 87,114 acres of the Sonoran Desert. The park is named for the saguaro cactus. These giant desert sentries can reach 50 feet in height and weigh up to several tons. They favor rocky mountain slopes and are a magnet for bird life, including red-tailed hawks, cactus wrens, elf owls, and Gila woodpeckers. Terrain ranges from low desert scrub to woodland and high-country forest. Look for jackrabbits, collared lizards, kit foxes, coyotes, and piglike javelinas.

Laguna San Ignacio
Baja California

Drifting across a placid seawater lagoon, the hollow sounds, deep and forceful, become more noticeable during the night. The chorus of giant exhalations resonates in winter-cool desert air. They pulse into a hushed campsite like ancient music, pushing gently through the thin nylon tents of sleeping journeyers at the edge of Baja California's **Laguna San Ignacio**. ◆ *Shoo-ssss. Shwooosh.* A faint splash. The lap of tiny waves against rock and sand. Then again: *Shwooshhhh!* Some campers waken to locate the source and direction of the noises. There. From the waters off **Punta Piedra**. All around! The gray whales are breathing. Their immense bodies – females average 46 feet and almost 70,000 pounds – rise and fall effortlessly, bringing twin blowholes above the surface for that wondrous and familiar exchange of fresh oxygen for carbon dioxide. ◆ Here at the edge of **Desierto de Vizcaino**, thousands of square miles of broad, flat desert greet the cool Pacific Ocean like an open palm, then rise

Harvested almost to extinction, majestic gray whales have returned in numbers to Baja lagoons to raise their young and mate.

gently from wandering beaches to the knuckled summits of **Sierra San Pedro** and **Sierra San Francisco**. Despite being halfway down the length of Baja California's 800-mile-long bony finger, and only a long day's drive north of the Tropic of Cancer, Laguna San Ignacio can be chilly. ◆ Drawn by the whales' seeming calls, inquisitive campers stand at water's edge beneath a sky so strewn with stars that even favorite constellations known from childhood are difficult to locate. There. Again. *Whooshhh!* With dawn, listeners become watchers. Light blushes behind eastern mountains. The lagoon mirrors a vivid turquoise sky. Each whale exhalation is now accompanied by

A gray whale thrusts its immense barnacle-covered head above the waters of Laguna San Ignacio.

a distinctive, heart-shaped 15-foot-tall vapor plume. A peaceful place, this San Ignacio. No wonder the whales have been coming here for many thousands of generations or more.

Close Call for the Gray Whale

Gray whales are the last living members of the baleen whale family known as Eschrichtiidae. They have survived for 20 million years in part because of their unique migratory habits. The whales have long journeyed 10,000 miles round-trip each year from the Chukchi, Beaufort, and Bering seas to coastal lagoons in Baja California. Protected by shifting bars and waves at the entrances, as well as hundreds of square miles of relatively shallow (60 feet or less) habitat, gray whales were mostly free from their only predators: sharks, killer whales, and, until 1857, humans.

Charles Melville Scammon, coasting in his whaler *Boston* along Baja California within the 70-mile-wide hook of Vizcaino Bay in 1857, had heard reports of whales hiding in undiscovered bays and estuaries. A sharp-eyed lookout atop the mainmast spotted several spouts seeming to come from behind a sand dune, and soon the slaughter was on, not only in **Laguna Ojo de Liebre**, the first

Nesting birds (left) such as this red-tailed hawk are not hard to spot in Baja, where vegetation is sparse.

Whales (below), perhaps drawn to an engine's sonic frequency, approach boats and allow watchers to reach out and touch their skin.

lagoon to be discovered, but, within a few seasons, farther south at Laguna San Ignacio and **Bahia Magdalena** as well. Once rendered, each whale yielded about 20 barrels of oil. Blood and offal soaked the beaches of Baja's whale sanctuaries, and a great stench of death settled over what had been a place of winter rest, birthing, and mating.

By the late 1800s, the gray whales were all but extinct. Forgotten, the remaining few hundred slowly restored the population. Then an age of motorized ships rediscovered this whale's close-to-the-coast habits, leading to another killing frenzy in the 1930s. After an outcry, harvesting of the Eastern Pacific herd of gray whales was banned altogether in North America in 1938. Today, the gray

whale has returned to, or possibly even exceeded, prewhaling numbers: more than 23,000 now migrate each year.

The same robustness that gives the gray whale its scientific name (Eschrichtius robustus), as well as the strength to dash a whaler's longboat, helps account for its comeback. Unlike the two other baleen families – Balaenidae (right whales) and Balaenopteridae (rorqual whales) – gray whales feed mostly on the bottom, powerfully plowing up beds of silt, turning to one side and sucking in a huge mouthful, then expelling mud and water through matted rows of fibrous baleen, leaving behind invertebrates such as amphipods. In summer feeding grounds, a full-sized gray whale may consume 2,400 pounds each day. Its stomach is capable of holding 660 pounds of food at one time. Nursing and recently weaned whales grow rapidly. A whale named "J.J.2" was rescued in 1997 as a dehydrated 14-foot, 1,500-pound calf beached at Marina del Rey, California. By her 10th month, J.J. was growing at a rate of nearly three pounds an hour, and a quarter inch each day, during her sojourn at Sea World in San Diego before being released in March 1998.

In Baja lagoons, whales occasionally exhibit feeding behavior, rising to the surface with bottom muck trailing from the mouth. Mothers may also be teaching their calves, born in December or January, the basics of foraging. Although the whales feed little here, these whale-nursery lagoons don't lack vigorous and exciting activity. The spectacle now attracts thousands of visitors a year to all three lagoons in Baja California.

Flukeprints and Spyhops

Onshore at San Ignacio's **Rocky Point** whale-watching camp, seven whale watchers, a naturalist, and a Mexican guide gather by a 22-foot Mexican skiff called a *panga.* Life jackets, tightened especially snug by first-time watchers who nervously sense the whales' great size and power, ride a little high up near the ears. Cameras and binoculars swing from necks. Hats and polarized sunglasses shade eyes from one of the greatest

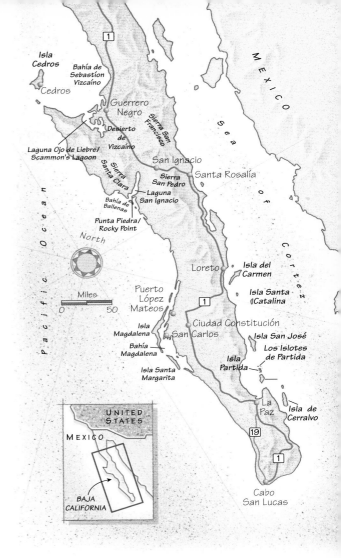

onslaughts of maximum lux anywhere in the world, and exposed skin is coated with waterproof sunblock.

The *panguero,* right hand on the tiller, knees slightly bent, stands beside his burbling outboard motor while whale watchers step aboard at the bow and take their places inside the fiberglass skiff. Remarkably sturdy and seaworthy, the 22-foot *pangas* can withstand the fulcral insult of having almost all passengers shift to one side in their eagerness to see a nearby whale. Everyone who wants to see a whale in Baja California up close eventually winds up in a *panga.* Although kayaking is allowed in some Mexican lagoons (most kayak trips take place

three-fourths of the way clear of the surface, then turning and falling onto their sides or backs with a mighty splash. This behavior happens regularly in the lagoons. One breach never seems to be enough: two or three times in succession is common.

Mating is also common, and two males are often seen approaching single females. A great rolling and thrashing may take place just beneath the surface, sometimes in water shallow enough that clouds of sand and silt kick up and boil to the surface. At times a male whale rolls over on its back, exposing what has come to be known in Baja California as its "Pink Floyd," a member thought to be able to move in search of the female genitalia, freeing the male from some of the logistics of maneuvering his entire bulk into perfect position.

The most anthropomorphically "cute" behavior may be seen in the mother's relationship with her newborn. Cow and calf maintain constant contact during the early days after a flukes-first, live birth. Mammary glands are hidden within folds in the mother's streamlined body. Calves may be nudged clear of the water for their first breaths by an "attending" male or female; later, for unknown reasons, mothers often lift calves on their backs until they roll off. With each passing week, the calves become increasingly independent, even approaching a *panga* and its whale watchers after their mothers have clearly turned away to leave.

Close Encounters

These "friendly" instances of whales approaching *pangas* (and people) for a pat or rub, and sometimes even allowing humans to touch their baleen, have had whale-watching aficionados abuzz since the behavior first occurred in 1976. Today, friendly encounters are most common at **San Ignacio**, but also take place at Magdalena Bay and **Scammon's**

in Magdalena Bay), the relatively tiny, one- or two-person craft are never permitted in active whale areas.

Soon the *panga* is motoring out into the main channel. Whales are everywhere, evidenced by their spouts and "flukeprints" – calm slicklike upwellings amid the choppy waves, caused by each stroke of their huge flukes beneath the surface. Knuckled dorsal humps curve and disappear. Suddenly, a mottled, barnacle-encrusted rostrum pops straight up out of the water near the boat as an adult female "spyhops" for several seconds, seemingly watching the watchers before slipping back under. Above water, whales are markedly nearsighted.

Spyhopping is fascinating behavior, carried out with powerful but subtle motions of flukes and pectoral fins, and some naturalists/cetologists who spend several months of every year at the whale lagoons think that the flukes might actually touch or push off bottom at times during spyhopping. Breaching may be part of courtship, or a method to shake loose barnacles, or a reaction to stress and predation, or simply something that feels good. Gray whales can launch themselves upward into the air with just two or three powerful kicks of their flukes, breaching

Lagoon (Ojo de Liebre). Wherever it happens, friendly behavior is initiated by a relatively few whales that are attracted to an idling boat engine's sonic frequency, a rub against the hull, the touch of humans, or possibly all of these factors and more. Regulations require *panga* captains to stay a good distance from any whale, but if a whale approaches, contact can begin. Passengers squeal, chatter, or sit silently in a state of wonder. Hats fly off in a sudden *whoosh* from the twin blowholes beside the gunwale. The whale's curved inverted "grin" is sometimes offered to several hands, the mouth opening slightly to expose baleen.

Close encounters also allow whale watchers to see the scabrous-looking barnacles and cyamid crustaceans or "whale lice" (possibly a symbiotic cleanup crew feeding on dying skin around the barnacles) on the dorsal surface and pectoral flippers of adult gray whales. Grays are the only whale so extensively infested with these organisms. The skin of the gray whale flakes and peels beneath skin adhesions on adults and overall on rapidly growing calves. Distinctive skin blotching, perhaps the result of hitchhiking barnacles come and gone, turns each whale into a white-gray-and-pink patchwork of scars, blotches, and bumps.

All the pats, strokes, even kisses may sometimes run contrary to naturalists' insistence on hushed observation. In truth, whale watchers mostly remember their sudden start at staring into a whale's eye for the first time. Bathed in seawater, it gazes into our world with a calmness we can only hope to find in our own species.

The Great Migration

The French call a long walk after a good dinner a *promenade digestif*. For the gray whale, blubber-wrapped and strong after a summer of bottom-feeding in the amphipod-rich silts of the Chukchi and Bering Seas between Alaska and Siberia, a 10,000-mile swim to the shallow lagoons of Baja California is far more than the oceanic equivalent of a stroll.

The imperative to find protected warm waters for mating and calving compels some 23,000 gray whales to make the astonishing round-trip every year between October and April. Theirs is the longest known migration of any mammal. As arctic waters cool and the ice pack once again stretches southward, the whales push through the Bering Strait, then unerringly find Unimak Pass, the first gap in the Aleutian Islands' curving barrier. Except for open-ocean stretches in the Gulf of Alaska and the Southern California Bight, the whales are often within sight of land from Canada to Baja, navigating by a combination of factors, especially water depth, currents, temperature, and vocalization.

Because of the great distance, many pregnant females are in the first wave southward, clipping off the miles with a stately urgency, swimming 20 hours and 100 miles per day. Whale watchers delight in seeing the whales' familiar respiratory cycle, particularly on the northbound leg when they stay much closer to shore: three to five spouts (exhalations) at the surface about 40 seconds apart, followed by a showing of flukes and a 10- to 20-fathom dive that covers perhaps another one-fifth of a mile. Whales usually swim singly, in pairs, or in trios, although groups of up to a dozen are occasionally seen. Males and in-season females may mate throughout the trip. By January, three huge lagoons on the west coast of Baja California are aboil with gray whales in their winter habitat.

Bottlenose dolphins (opposite, above) travel in groups, communicate using low-frequency sounds, and feed cooperatively by herding schools of fish.

Brown pelicans (opposite, below), with their seven-foot wingspan, seem ungainly but are expert divers that capture fish in their pouches.

Tail flukes (above), measuring 10 feet across, are a whale's primary means of propulsion. Using their powerful flukes, grays spyhop to view their surroundings.

TRAVEL TIPS

DETAILS

When to Go

Temperatures on the Pacific Coast hover around the mid-80s most of the year, somewhat cooler in winter. Sunshine is plentiful, rain unusual. Traveling in the interior is another matter. The desert can be brutally hot and dry, with summer temperatures routinely soaring above 110°F. Be sure to bring sunglasses, a hat, and plenty of sunscreen.

How to Get There

Airports with international connections are based in Tijuana, Mexicali, Loreto, La Paz, Los Cabos, and San Jose del Cabos. Several companies offer tours and transportation from San Diego (see Tours & Outfitters).

Getting Around

Car rentals are available at all Baja airports. An automobile is the most convenient way to travel, but information about long-distance bus service in Baja California is available from Tijuana's Central de Autobuses, 52-66-86-9060. Vehicles with four-wheel-drive may be helpful on remote, unpaved roads.

INFORMATION

Baja Information

7860 Mission Center Court No. 2; San Diego, CA 92108; tel: 800-225-2786 or 800-522-1516 (in California).

CAMPING

Baja Discovery

P.O. Box 152527; San Diego, CA 92195; tel: 800-829-2252 or 619-262-0700.

Rocky Point Camp is situated on a low bluff overlooking San Ignacio Lagoon, providing round-the-clock whale-watching opportunities. Visitors stay in roomy tents with comfortable cots. The camp has modern bathroom facilities and solar showers. Meals, served in a large dining and activity tent, are a cut above ordinary camp food.

El Padrino Trailer Park

San Ignacio 23930; tel: 115-4-00-89.

Located on the main road into San Ignacio, across from Hotel La Pinta, this park has 10 spots for recreational vehicles and 40 campsites in a grove of date palms. Large grills, barbecue pits, a restaurant, and bar are available.

LODGING

PRICE GUIDE – double occupancy

$ = up to $49 $$ = $50-$99
$$$ = $100-$149 $$$$ = $150+

Hotel Blanco y Negro

Avenida Sarabia 1; Santa Rosalia; tel: 115-2-00-80.

Basic but comfortable lodgings with reliable hot water; some with private bath. A spiral staircase leads to 12 guest rooms on the second floor. $

Hotel El Morro

Apdo Postal 76; Santa Rosalia; tel: 115-2-04-14.

This Spanish-style hotel overlooks the Gulf of California a mile south of the Santa Rosalia ferry terminal. Rooms have two beds and private baths; some have private patios. Amenities include a pool, dining room, and bar. $

Hotel La Pinta

P.O. Box 37; San Ignacio 23930; tel: 800-336-5454.

Located on the main road to San Ignacio, this Colonial-style hotel offers comparatively fine accommodations, including a tiled courtyard, swimming pool, and groves of date palms and various citrus trees. $$

Hotel del Real

Avenida Montoya; Santa Rosalia; tel: 115-2-00-68.

An attractive wooden building with a terrace, good restaurant, and long-distance phone. Rooms are small but air-conditioned. $

Hotel Frances

Avenida Cousteau 15; Santa Rosalia; tel: 115-2-20-52.

The hotel offers simple but comfortable lodging in 17 air-conditioned rooms. Swimming pool. $

Motel La Posada

Avenida Carranza 22; San Ignacio 23930; tel: 115-4-03-13.

Small, Spartan, inexpensive accommodations, with double beds, fans, and hot showers. The motel offers tours to the lagoon and nearby pictographs. $

TOURS & OUTFITTERS

Baja California Tours

7734 Herschel Avenue, Suite O; La Jolla, CA 92037; tel: 619-454-7166.

A seven-day "Follow the Whales" tour departs from San Diego. Deluxe air-conditioned motor coach and air transportation provided. Passengers stay in hotels in Catavina, San Ignacio, Loreto, Puerto San Carlos, and La Paz, and enjoy whale watching from 22-foot skiffs in Scammon's Lagoon, Puerto Lopez Mateos, and Puerto San Carlos.

Baja Discovery

P.O. Box 152527; San Diego, CA 92195; tel: 800-829-2252 or 619-262-0700.

Expert guides and naturalists lead tours to the Pacific Coast and Sea of Cortez. Five- to eight-day whale-watching trips combine hotel stays and beach camping. Other tours focus on birding, botany, cave paintings, and hiking.

Baja Expeditions

2625 Garnet Avenue; San Diego, CA 92109; tel: 800-843-6967 or 619-581-3311.

In operation since 1975, "Mexico's largest and oldest outfitter of adventure travel" offers flexible itineraries for whale watching, sea kayaking, sailing, and scuba diving. Whale-watching trips in the Sea of Cortez may encounter humpback, sperm, finback, minke, gray, blue, and killer whales, as well as dolphins. Frequent stops are made on desert islands for beachcombing, snorkeling, and hiking.

Baja Quest

Sonora #174; La Paz, 23060; tel: 112-353-20.

Whale watching, sea kayaking, and camping in and around Magdalena Bay. Island-based camps offer easy motorized skiff or kayak access to prime gray-whale areas. Birding is excellent, too. Four- and five-night itineraries combine camping and hotel lodging. Trips also include ground transportation, naturalist guides, and meals.

Tio Sports

P.O. Box 37; Cabo San Lucas; tel: 114-3-33-99.

Fully equipped scuba diving, snorkeling, sea kayaking, horseback riding, and mountain biking in Los Cabos.

Excursions

Bahia Magdalena

"Mag Bay" to many English-speakers, Bahia Magdalena is the gateway to the Pacific. Gray whales arrive from January to March, transforming the area into one of the coast's most important whale-breeding sites. The 50-mile-long bay is protected from strong Pacific waves by Magdalena and Margarita Islands. Beaches, inlets, marshes, and mangrove swamps sustain a variety of resident and migratory seabirds.

Laguna Ojo de Liebre

Ojo de Liebre, also known as Scammon's Lagoon, is the largest gathering place of California gray whales, which come each spring for birthing and mating. Mexico created the world's first sanctuary for gray whale here in 1972.

Birds, also sheltered by the lagoon, include Canada geese, ospreys, cormorants, and white pelicans. Unfortunately, the lagoon is also the site of the world's largest evaporative salt works, which many environmentalists fear will have detrimental effects on this vital habitat.

Parque Nacional Sierra San Pedro Martir

KM 107 Carretera, Tijuana-Ensenada; Ensenada 22860; tel: 52-61-74-08-88.

Established in 1947, this 236-square-mile park, located in the center of northern Baja, preserves a variety of terrain, vegetation, and animal life. Many granite peaks, such as the impressive Picacho del Diablo, exceed 10,000 feet. Jeffrey and Parry pines fill out the park's beautiful forest. Conspicuous fauna includes coyotes, mule deer, gray foxes, and raccoons. Mountain lions are occasionally spotted, as are rare desert bighorn sheep. Birds include pinyon jays, pygmy nuthatches, Cassin's finches, and red crossbills. The park also is inhabited by many species of amphibians, reptiles, and fishes.

Monterey Bay
National Marine Sanctuary
California

CHAPTER **17**

A forest of giant kelp lies 40 feet deep at the mouth of **Whalers Cove**, swaying to the hypnotic rhythm of the Pacific Ocean. Sunlight streams from cracks in the fast-growing kelp canopy carpeting the surface. Sponges and fanlike algae coat the submerged boulders in shades of lavender, with splashes of orange, scarlet, and cobalt blue. As one sponge moves and sprouts spidery, clawed legs, you see that it and several bits of red algae help disguise a masking crab. Tucked under a nearby ledge, your diving buddy points out a cotton-candy-colored sunflower star, two feet across, with arms that stick like Velcro to your neoprene gloves. ◆ All around, rockfish swim placidly or lurk in rock crevices, unafraid of the rubber-clad creature with bubbles rising behind its head. A ghostly apparition cruises through the kelp, causing your bubble trail to halt for an instant as the image of a great white shark – top predator in these waters – pops into your brain. But it's just a mottled gray harbor seal, and you resume breathing steadily. ◆ Apart from the remote chance of encountering a white shark, **Whalers Cove** in **Point Lobos State Reserve** looks and feels like a sanctuary. Fish are tame and abundant. Curious sea lions come close and perform pinniped pirouettes. Sea otters forage freely for urchins and abalone and devour them on their backs, oblivious to scuba divers surfacing a few yards away. Only 15 teams of divers a day are allowed in Whalers Cove or neighboring **Bluefish Cove**, and all catching and collecting are prohibited. In 1960, this portion of the park became the country's first underwater reserve.

An abundance of creatures inhabits this sanctuary's kelp forests, tide pools, salt marshes, and pine and cypress woodlands.

Bat stars nestle among plants in the intertidal zone of Monterey Bay, which is at the center of a 300-mile-long underwater refuge.

Web of Life

These waters belong to a much grander sanctuary, one that extends into the ocean for up to 53 miles and north and south along one-fifth of the California coast. **Monterey Bay National Marine Sanctuary** comprises the largest link in the nation's chain of 12 areas deemed the most significant natural spots in our seas. Established in 1992, it stretches 300 miles from Cambria in the south to the headlands north of the Golden Gate Bridge, protecting more than 5,200 square miles centered on crescent-shaped Monterey Bay.

Among the sanctuary's stars are animals rebounding from brushes with extinction: the northern elephant seal, California gray whale, brown pelican, and southern sea otter. Visitors may soon glimpse wild California condors soaring along this coast. The sanctuary's diverse wildlife includes 26 species of marine mammals, 94 seabirds, and 345 fish, plus an incredible richness of invertebrates: anemones, corals, crabs, chitins, seastars, snails, sponges, and squid. Such bounty occurs here because the largest submarine

canyon in the continental United States lies offshore and drops to a depth of 10,000 feet. The seaward diversion of surface currents by the Earth's rotation causes cold, nutrient-rich water to rise from the deep (a process called upwelling) and fuel the food web that links animals as tiny as plankton and as mammoth as whales.

Upwelling means that divers, too, must tolerate frigid water, around 50°F, or visibility that can be next to nothing. At its best, though, on a clear autumn afternoon when underwater visibility can exceed 60 feet, a dive in a Monterey kelp bed rivals one at any tropical reef. Though this coast has been protected for its underwater wealth of resources, with dozens of great scuba sites close to shore or reachable by boat, wildlife watchers can experience the sanctuary without donning a wet suit – even without getting wet. Outposts accessible on foot or by kayak abound along this coast, and boat tours that go in quest of seabirds and cetaceans depart Monterey or San Francisco for the open ocean.

Sea Otters

Once hunted nearly to extinction for its incredibly lush fur – a million hairs per square inch – but now protected, the southern sea otter is reclaiming old haunts along California's coast. From shore, it can be tough to spot this smallest marine mammal atop a bed of bobbing kelp. Look for a white head and hind flippers, or listen for an otter smacking shellfish against a flat rock on its belly. Sea otters must eat lots of food to heat their blubberless bodies. After a meal, they wash off by spinning around in the water. Curious and clever, otters have been known to hitch rides on passing kayaks, harvest oysters with a bottle, and play with sunken cans, golf balls, and tires.

An anemone (above) waving its tentacles, its mouth open wide, waits to disable and consume passing fish.

Sea otters (left) use their bellies as a platform for extracting the flesh from crabs, urchins, and abalone. Once hunted for their fur, they are protected and making a comeback.

Onshore Forays

Landlubbers can sample the marine riches by exploring tide pools at **Fitzgerald Marine Reserve** near **Half Moon Bay**. Low tide at the reserve reveals California's largest accessible reef. Life is harsh here in the inter-tidal zone, where plants and animals endure a constant cycle of exposure and inundation by the ever retreating and advancing ocean. Each tide pool in the ancient shale contains a miniature marine sanctuary: anemones in pastel reds and greens, foot-long gumboot chitons, odd-looking sea cucumbers, and sculpins darting in and out of nooks.

Should you tire of searching for the animals, try identifying the seaweeds. About 700 different kinds can be found in this 40-acre reserve. Other prime tide-pooling locations within the national sanctuary include **Point Pinos** in **Pacific Grove** and **Weston Beach** at **Point Lobos**. Be sure to consult tide tables for the time and extent of low tide, wear sturdy shoes, and try not to trample the local residents.

Point Lobos, four miles south of **Carmel**, offers almost all the sanctuary's wonders in microcosm. Arrive early to stay ahead of the crowds lured to this gorgeous stretch of central coast scenery. The **North Shore Trail** from the **Whalers Cove** parking area provides an excellent introduction. You won't get far without meeting plump Beechey ground squirrels on the lookout for handouts. Linger on the bluff overlooking Whalers Cove, a good place to spot otters in the kelp or harbor seals curled atop rocks. In summer, poison oak sprouts warning colors and the hillsides blaze with apricot blossoms of sticky monkeyflower. Northern flickers nest in Monterey pine snags draped in wispy, pale-green lichen. A covey of valley quail – parents with a dozen fuzzy chicks – flush from a trailside tangle of undergrowth and vanish just as suddenly among cypress branches.

The trail passes high above **Bluefish Cove**, where a great blue heron or snowy egret might be perched on a drift log, its head cocked toward the water. Scan the rocks in Bluefish for more flippered flotsam – lounging seals. Beneath the surface,

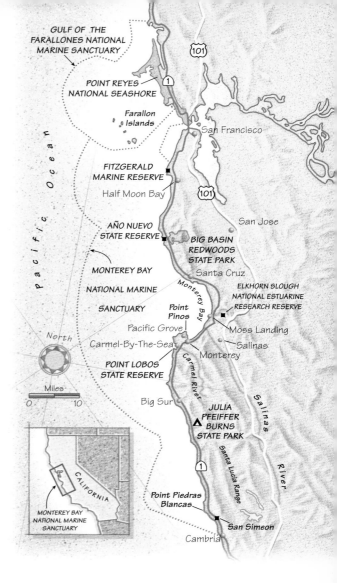

Bluefish Cove contains the reserve's richest marine life in a maze of rocky canyons, but divers will need a boat or kayak to get here. Leaving a cover of Monterey pine, you enter a fairy-tale grove of gnarled and wind-whipped Monterey cypress. Only at Point Lobos do these two trees grow side by side. Walk to the overlook for a view of **Guillemot Island**, a guano-streaked granite monolith frequented by gulls, cormorants, and pigeon guillemots.

The short **Sea Lion Point Trail** leads to the prominence that lends its name to the reserve, **Punta de los Lobos Marinos** (Sea Wolf Point). But rather than howl like wolves, the sea lions honk and bark in a constant chorus and frolic in the water, raising noses and flippers into the air. In summer, immature males hang out here, and by fall adult males begin to arrive from breeding areas farther south. In late spring, this point makes a prime place for spotting mother gray whales migrating north with newborn calves in tow.

Salt Marshes and Tidal Flats

Midway along the arc of Monterey Bay, **Elkhorn Slough** makes a memorable stop for wildlife watchers. The intrepid can slip into a sea kayak to explore the slough's main channel, which meanders inland seven miles along the sanctuary's largest salt marsh, more than 3,000 acres. An organized tour by flat-bottomed boat from **Moss Landing** offers another option. On the north side of the slough, east of Highway 1, the **Moss Landing Wildlife Area** has a marsh-side hiking trail – take time to squat down and crunch a salty stem of pickleweed, the predominant plant – that leads to a picnic area. A wildlife blind at the end of **West Trail** overlooks expansive mudflats being reclaimed by shorebirds and waterfowl from their prior use as commercial salt ponds. The mudflats are host to the largest summer roosting population of brown pelicans north of Point Conception, and shorebirds arrive in winter by the thousands.

At **Elkhorn Slough National Estuarine Research Reserve**, on the slough's south side, learn about the hundreds of unseen and unsung marsh mud dwellers, such as clams, shrimp, snails, and worms, from clever exhibits in the visitor center. From a nearby overlook, the slough shimmers in the sun between parched hills and irrigated fields. Two permanent spotting scopes, one set at wheelchair level, bring the scenery closer and entice you to set out on the five-mile network of easy unpaved trails.

Take a morning hike on the one-mile **Five Finger Loop Trail** from the wildlife blind nestled under a cluster of oaks. There, you can watch a willet dart across the mud to snatch an unwary crab in its bill and fly off as another bird tries to wrest away the catch. Marbled godwits, sandpipers, dow-itchers, and curlews scurry and probe the mud with beaks of various sizes and shapes. Terns glide back and forth, while egrets roost on leftover levees.

Along the **South Marsh Loop Trail** in June, you may glimpse leopard sharks or bat rays coming into the shallows and tidal creeks to bear their young. Great egrets fly

Kelp (opposite), a form of brown algae, rises from the ocean floor to the surface, forming the canopy of a dense underwater forest that provides shelter and food for marine creatures.

Humpback whales (right), as well as minke and blue whales, are seen in the waters of the refuge that also protects the Farallon Islands.

Seals and sea lions and their young (below) crowd the rocky shore.

Island Refuge

Twenty-seven miles west of the Golden Gate Bridge, jagged granite crags jut from the open ocean. Gulls circle and cry overhead. The stench of guano fills the air. Thousands of black-and-white common murres dot sea caves and cliff faces. The biggest U.S. seabird rookery outside Alaska, with 150,000 birds from 12 species, the Farallon National Wildlife Refuge harbors many of the world's western gulls, ashy storm petrels, and Brandt's cormorants.

Wildlife on the Farallones, California's northernmost islands, come and go with the seasons. Windy spring and summer bring abundant seabirds to nest on precarious cliffs or in burrows. Cormorants denude the island of its verdant cloak of Farallon weed to line their nests. Gull nests with mottled olive eggs or awkward chicks lie scattered every few feet. Humpback, blue, and minke whales feast in the plankton-rich waters in the Gulf of the Farallones National Marine Sanctuary. In fall, seals and sea lions haul ashore to bear pups, while great white sharks arrive to patrol the ocean for seal meals. With winter storms come brawling and breeding northern elephant seals and California gray whales migrating past to their nursery lagoons in Mexico.

Only a few biologists who occupy the island year-round can set foot on Southeast Farallon or the smaller islands, but tourist boats leave San Francisco for day-long nature cruises to the Farallones in summer and fall. Although rough water during the voyage can turn even an experienced sailor queasy, the journey provides a glimpse of seabird and marine life. Auklets and murres skim the sea surface with anchovies dangling from their bills. About 400 bird species have been identified at the islands, including a few oddities blown far off course. The onboard naturalists won't guarantee whale sightings, but encounters occur often with grays and humpbacks, plus bow-riding dolphins and porpoises.

over the trail between the slough and their nests. The trail passes close by their rookery in a grove of Monterey pines beside a pond. Several dozen pairs of egrets and fewer great blue herons build bare twig nests here each spring. Double-crested cormorants have recently joined the rookery and nest at the very tops of the trees. By late summer, the nearly fledged egrets still start a commotion, flap their wings, and grasp an adult's bill after the parents swoop in from

another foraging trip. A short distance from the rookery pond, watch for acorn woodpecker granary trees – snags riddled with holes – and the red-capped, black-and-white birds themselves.

A Bounty of Seals

At **Año Nuevo State Reserve**, 18 miles north of **Santa Cruz**, you can witness a conservation comeback as thousands of northern elephant seals hit the beaches at

the point and on the adjacent island. Bull males, the largest seal on North American shores, arrive in winter to contest for mates in violent bouts of biting and body slamming. Then females come ashore to give birth, nurse the newborns for a month, and mate again before heading back to sea to restore their body mass during feeding dives thousands of feet deep.

Earlier in the 20th century, northern elephant seals had been hunted to the last few hundred. From a safe haven on Guadalupe Island in Mexico, they steadily expanded their range and numbers along the West Coast to more than 150,000 today. The seals first occupied **Año Nuevo Island**, also a stop-over for California sea lions, which take over the abandoned lighthouse keeper's dwelling, and rare Steller sea lions. Soon elephant seals were spilling over to the mainland, where the first pup was born in 1975. Now about 200,000 people show up each year to see the thriving colony.

From December through March, visitors must join a guided tour of the breeding grounds, and it's wise to book ahead. December features the bouting bulls, and January is a peak month for new pups. The numbers peak in spring when juveniles and adults return to molt. Stacked like cord-wood along the shore, the seal colony emits a cacophony of snorts, sighs, belches, and bellows, punctuated by the percussive riffs from a male's pendulous nose. After March, visitors can obtain a permit to visit the dunes without a guide but still must stay 20 feet from any seal. During the walk to the dunes, watch for northern harriers on the hunt, flying just above the dense scrub. And keep an eye out for the endangered San Francisco garter snake, a stunning reptile with brick-red and sky-blue stripes. At the southern end of the Monterey Bay sanctuary, the state's newest and fastest-growing elephant seal colony continues to annex beaches south of **Point Piedras Blancas**. A single pup was born there in 1992; five years later, females gave birth to 1,200 pups. Now the colony numbers more than 4,000 seals. Observe the seals from vista points above the beach – it's illegal and unsafe to get close – and heed the yellow traffic signs that caution highway drivers to watch out for stray seals.

Whether you stick to shore or choose to venture on or under the water, you will be rewarded, and astounded, by time spent exploring the Monterey Bay National Marine Sanctuary. From modest mud invertebrates to the cutest and most charismatic mammals, wildlife flourishes throughout the region. Dive right in.

California sea lions (left) and elephant seals take over stretches of the coast near Monterey Bay, drawing many visitors from late winter through spring.

Western gulls (above) undergo changes in plumage through the seasons and as they mature, making them difficult to identify.

Wildlife watchers who don't dive have plenty to discover in tide pools, from anemones and hermit crabs (right) to sea cucumbers and fast-moving sculpins.

TRAVEL TIPS

DETAILS

When to Go

Weather is mild year-round. Summer is warm and sunny; winter is cool, blustery, and rainy. Spring and fall have the most pleasant conditions, with temperatures in the 50s and 60s. Whale watching is best during the spring and winter migration of California gray whales. Elephant seals give birth and mate November through March.

How to Get There

Monterey Peninsula Airport, a hub for a number of major airlines, is located three miles from downtown Monterey. A three- to five-hour bus ride between San Francisco and Monterey is made by Greyhound Lines, 800-231-2222. Amtrak's Coast Starlight train, 800-872-7245, stops in Salinas en route from Los Angeles to Seattle.

Getting Around

Car rentals are available at the airport. Automobiles are essential for travel in the Monterey Bay area and between parks. Whale-watching and sightseeing boats are available at Fisherman's Wharf in Monterey.

INFORMATION

Monterey Peninsula Visitors and Convention Bureau

380 Alvarado Street; Monterey, CA 93942; tel: 831-649-1770.

California State Division of Tourism

801 K Street, Suite 1600; Sacramento, CA 95814; tel: 800-462-2543 or 916-322-2881.

CAMPING

Los Padres National Forest and many state beaches and parks provide hundreds of campsites in Monterey County.

LODGING

PRICE GUIDE – double occupancy

$ = up to $49 $$ = $50-$99
$$$ = $100-$149 $$$$ = $150+

Cobblestone Inn

P.O. Box 3185; Carmel, CA 93921; tel: 800-833-8836 or 831-625-5222.

Recently renovated, this former motel is now an English-style inn, complete with stone fireplaces in the guest rooms and sitting room. The inn's 24 rooms are furnished with country antiques, queen- or king-sized beds, and a refrigerator. $$–$$$$

Mission Ranch

26270 Dolores Street; Carmel-by-the-Sea, CA 93923; tel: 800-538-8221 or 831-624-6436.

This 1850s farmhouse was salvaged by Clint Eastwood. The ranch complex has 31 rooms, several of which offer ocean views, fireplaces, Jacuzzi tubs, stuffed mattresses, and carved wooden beds. A restaurant with panoramic views, a piano bar, an exercise room, tennis courts, and a putting green are on the grounds. $$–$$$$

Monterey Hotel

406 Alvarado Street; Monterey, CA 93940; tel: 800-727-0960.

Built in 1904, this downtown hotel reopened in 1987 and was renovated in 1996. Victorian touches include oak paneling, ornate fireplaces, marble floors, and antique furnishings. The hotel's 45 rooms have private baths; suites have sunken baths and fireplaces. $$$–$$$$

Old St. Angela Inn

321 Central; Pacific Grove, CA 93950; tel: 800-748-6306 or 831-372-3246.

This Craftsman-style Cape Cod house was built in 1910 and later converted to a rectory, then a convent. Overlooking Monterey Bay, the inn is a block away from rugged coastline and within walking distance of the Monterey Bay Aquarium. Eight guest rooms have private baths, many with whirlpool and ocean view, some with fireplace. The house contains antique furniture, a stone fireplace, and a solarium. $$–$$$

Pine Inn

P.O. Box 250; Carmel-by-the-Sea, CA 93921; tel: 800-228-3851 or 831-624-3851.

Just four blocks from the sea, this popular inn was built in 1889 and still maintains a high standard of elegance. Forty-nine guest rooms offer many options, including suites, ocean views, and both European and Far Eastern styles. The lobby's Victorian decor features handsome wood paneling and fine antiques. Specialty shops and a restaurant are on the premises. $$–$$$$

TOURS & OUTFITTERS

Elkhorn Slough Safari

P.O. Box 570; Moss Landing, CA 95039; tel: 831-633-5555.

Cruises on a 27-foot pontoon boat allow passengers to observe wildlife, such as otters, seals, waterfowl, and migratory shorebirds. Reservations should be made at least one week in advance.

Monterey Bay Kayaks

693 Del Monte Avenue; Monterey, CA 93940; tel: 800-649-5357.

Natural-history kayak tours for expert and beginner paddlers in Monterey Bay National Marine Sanctuary and Elkhorn Slough National Sanctuary. Naturalists guide paddlers to views of sea otters, sea lions, and many birds.

Monterey Bay Whale and Nature Cruises

P.O. Box 52001; Pacific Grove, CA 93950; tel: 831-375-4658.

Whales, dolphins, seals, sea lions, otters, and seabirds are frequently seen during these boat trips, which last up to six hours. Trips depart from Fisherman's Wharf in Monterey.

Pebble Beach Equestrian Center

Portola Road and Alva Lane; P.O. Box 1498; Pebble Beach, CA 93953; tel: 831-624-2756.

Instructors lead horseback rides to Del Monte Forest, the Pacific coastline, and other scenic spots. Riders sometimes spot deer, foxes, whales, sea lions, and sea otters.

MUSEUMS

Monterey Bay Aquarium

886 Cannery Row; Monterey, CA 93942; tel: 831-648-4888 or 800-756-3737 (reservations).

State-of-the-art exhibits include the million-gallon Outer Bay wing, a three-story kelp forest, a walk-through aviary, and touch pools.

PARKS

Año Nuevo State Reserve

New Year's Creek Road; Pescadero, CA 94060; tel: 650-879-2025.

Elkhorn Slough National Estuarine Research Reserve

1700 Elkhorn Road; Watsonville, CA 95076; tel: 831-728-5939.

Gulf of Farallones National Marine Sanctuary

Fort Mason, Building 201; San Francisco, CA 94123; tel: 415-561-6622.

Monterey Bay National Marine Sanctuary

299 Foam Street; Monterey, CA 93923; tel: 831-647-4201.

Point Lobos State Reserve

Route 1, Box 62; Carmel, CA 93923; tel: 831-624-4909.

Excursions

Channel Islands National Park

1901 Spinnaker Drive; Ventura, CA 93001; tel: 805-658-5700.

The United Nations designated this park an International Biosphere Reserve. Created in 1980, the park encompasses five islands and the surrounding sea and is accessed by boat and small plane only. Channel Islands protects many rare and threatened species of plants and animals, as well as a giant kelp forest with about a thousand kinds of fish and plants. Wildlife includes whales, dolphins, sea lions, 20 species of sharks, and a great abundance of bird life.

Golden Gate National Recreation Area

Building 201, Fort Mason; San Francisco, CA 94123; tel: 415-556-0560.

The largest urban national park in the world encompasses 76,500 acres of land and water, includes 28 miles of coastline, and is nearly two and a half times the size of San Francisco. Extending north of the Golden Gate Bridge along the Marin Headlands, the park protects bays, beaches, grasslands, forest, and 2,600-foot Mount Tamalpais. Towering coast redwoods distinguish Muir Woods, while many species of waterfowl and seabirds avail themselves of the park's lovely streams, ponds, lagoons, and marshes. Mule deer, badgers, bobcats, and gray foxes are but a few of Golden Gate's wild denizens.

Point Reyes National Seashore

Point Reyes, CA 94956; tel: 415-663-1092.

This extraordinary park just 20 miles north of San Francisco preserves 86,000 acres of grasslands, forest, windswept coast, and estuaries. Visitors can witness tule elk bulls sparring during the late summer and early fall, or catch sight of fallow and axis deer – both exotic species – and native black-tailed deer. Birding is excellent throughout the park. More than 430 bird species have been spotted here, and thousands of raptors migrate through in late fall and early winter. Overlooks on the coast are good spots to observe sea lions, elephant seals, shorebirds, and migrating gray whales.

Yosemite
National Park
California

CHAPTER **18**

For anyone who has seen an Ansel Adams photograph of Half Dome or El Capitan, Yosemite engenders an attraction so strong that it can become a physical yearning. The great granite walls rising from an arcadian landscape of meadow and forest, the rushing rivers and ethereal waterfalls, are somehow Edenic, somehow perfect, without flaw. Yosemite Valley looks so poignant and familiar that it seems stamped on our genetic memory. ◆ Yet the park is more than this single stunning valley. A pristine 1,169-square-mile swatch of the Central Sierra, **Yosemite National Park** is home to an astonishing array of animals. Wildlife thrives in Yosemite, both in the valley and in the backcountry, but most of the park's wild species are retiring, and it takes patience and some effort to find them. To fully appreciate the park, then, drinking in the stunning vistas isn't enough. You have to look closely and move quietly. The wildlife is all around you; to see much of it, you need to become part of the landscape. ◆ Yosemite's varying

Soaring golden eagles, bold black bears, and elusive mountain lions haunt the pristine Sierra Nevada.

elevations, rugged topography, and abundant rivers and streams make it rich in ecotones – edge areas where different environments meet, creating a variety of habitats. The riparian zone is a particularly good place to explore, since it encompasses several distinct habitat niches along with abundant food, water, and cover. Two rivers dominate the park: the Merced and the Tuolumne. Both support an abundance of small mammals, including the gray fox, raccoon, striped skunk, spotted skunk, and bobcat, and perhaps even the exceedingly rare Sierra red fox. All of these animals are reclusive and

Yosemite Falls, a torrent in spring, feeds the Merced River as it flows through Yosemite Valley, forming a corridor that attracts deer, coyotes, and other wildlife.

Pacific
Crest Trail

Saddlebag
Lake

Tioga Crest

Hetch Hetchy
Reservoir

Grand Canyon of
the Tuolumne

Harden
Lake

Tuolumne River

Glen Aulin

Tioga Pass
Entrance

120

Tioga Pass Road

Big Oak Flat
Entrance

120

Tuolumne
Meadows
Visitor Center

Cathedral Range

Lyell Canyon

395

Valley
Visitor
Center

Crane Flat

El Capitan

Half Dome

Clark Range

El Portal

Yosemite
Valley

Glacier
Point

Merced River

Badger
Pass

Arch Rock
Entrance

Mariposa Grove

49

140

Wawona

South Entrance

140

41

CALIFORNIA

YOSEMITE
NATIONAL
PARK

too, like the valley, and are particularly fond of the meadowlands, where they hunt voles, gophers, and mice, often within proximity of alert wildlife watchers. Black bears regularly patrol the Merced and nearby campgrounds and parking lots, though rigorously enforced regulations and bear-proof garbage containers help minimize unpleasant encounters between bruins and visitors. Many of Yosemite's bears are more accurately described as feral than wild – they are not truly tame, obviously, but their natural instincts have been dulled by human contact and human food.

generally nocturnal, though they often leave their tracks and scat on sandbars and banks.

The **Merced River**, the better known of the two, wends through the lush meadows of **Yosemite Valley** before it exits the park and plunges down a long gorge. This portion of the park is bordered by a patchwork of grassland and stands of ponderosa pine, white fir, black oak, incense cedar, and cottonwood. Trails skirt virtually the entire riverway. Though Yosemite Valley gets most of the park's visitors, some wildlife species are relatively abundant – due partly to patches of good habitat that have survived along the Merced and partly to the simple fact that many animals have become habituated to human beings.

Most of the mammals are crepuscular, and the best way to see them is by strolling slowly along the more isolated portions of the Merced at dawn and dusk. Black-tailed deer, the Pacific Coast subspecies of the larger mule deer, commonly browse the meadows in small groups in spring and summer or forage for acorns under the oaks in late autumn. Coyotes,

Yosemite's other big predator is rarely seen in or around the valley and generally prefers to haunt the backcountry. Mountain lions have enjoyed a significant resurgence in California since the passage of a 1991 state initiative granting them full protection. At least 5,000 now inhabit the state, and the Central Sierra probably supports its maximum sustainable population of the big cats. These opportunistic hunters prefer larger prey such as deer and bighorn sheep. Observations are apt to be fleeting; tracks are more commonly found. It's hard to mistake the tracks of a mountain lion for those of any other animal. They have a round shape and are about four inches in length, with a rear pad and four toes. Claw marks are not visible, as is the case with canids. Lion kills are distinctive and are sometimes discovered by hikers. Any deer or bighorn carcass partially covered by dirt or vegetation is probably the work of a lion.

Yosemite's Grand Canyon

The **Tuolumne River** affords more of a true wilderness experience than the Merced simply because it sees far fewer visitors. Much of the Tuolumne that lies within the park cuts through a remarkable declivity known as the **Grand Canyon of the Tuolumne**, the upper portion of which can be reached in a vigorous day hike from **Tuolumne Meadows** on **Tioga Road**, 55 miles from Yosemite Valley. A short trail just west of the Tuolumne Meadows store connects with the **Pacific Crest Trail**, which wends northwest about four miles before it joins with the **Grand Canyon of the Tuolumne Trail** near **Glen Aulin**. Tioga Road also provides access to numerous other trailheads – almost 20 in all – for those who want alternative hikes.

The Grand Canyon of the "T," as kayakers and rafters refer to the river, is a wonderland of waterfalls, smooth Sierra granite shield, alpine conifers, and lush meadows. The wildlife here is suited for an alpine existence, inured to short summers and long winters. Yellow-bellied marmots, chunky relatives of the eastern woodchuck, inhabit scree slopes and boulder fields. They often stand sentinel on large rocks and chirp sharply when alarmed. Though they are wary of being approached directly, it is possible to get close to them by walking slowly on a course oblique to their position. Living in close association with marmots are pikas, small mammals that look like short-eared, truncated rabbits. Pikas usually live in colonies within rockfalls near grassy meadows. During the brief High Sierra summer, they industriously cut and dry grasses for hay, which they store in their dens for winter sustenance. These extremely timid animals are more often heard than seen. Their call, a distinctive, subdued bleat, frequently emanates from deep within rockfalls.

Wherever there are marmots, you'll also find their primary predators: golden eagles. These magnificent raptors swoop suddenly

Black bears (opposite), accustomed to human presence, freely wander into popular areas of the park such as Yosemite Valley.

Bobcats (right) prey on voles, mice, squirrels, and other small mammals.

over ridgetops to ambush dozing marmots, or wheel high overhead on thermals, scanning the terrain for prey. Their size makes them difficult to confuse with any other bird of prey, except for bald eagles. Young goldens and balds are similar, displaying mottled white and brown plumage on their under-sides and brown dorsal plumage.

Another bird of prey of enduring interest is the peregrine falcon. Several pairs nest in the park, and they are sometimes observed soaring near granite walls and cliffs. Sightings are fairly common along **Tioga Road** or on hikes along the **Pacific Crest Trail** through **Lyell Canyon**, south of **Tuolumne Meadows**.

Both the Tuolumne and the Merced are prime habitat for American dippers, known more poetically as water ouzels. These gray, robin-sized birds have the singular habit of diving into streams, where they stroll along the bottom, snapping up caddis larvae and other aquatic insects. They typically make dramatic entrances into fear-some rapids, then explode from the white water to perch, preen, and "dip" on stream-side boulders.

Yosemite's riparian woodlands – along the Tuolumne, Merced, and myriad other rivers and creeks – contain good songbird habitat. Late spring and early summer are prime times for birding, when warblers, thrushes, flycatchers, and other neotropical migrants are in their glorious breeding plumage. Western tanagers are a star species: there's no mistaking the male, resplendent

Clark's Nutcracker

Birds and beasts of the forest rarely let human beings approach close enough to observe them. One exception is the Clark's nutcracker. Found throughout the Sierra Nevada near timberline, these pigeon-sized, gray-and-white corvids delight in human company, or at least in the food they have learned to associate with campsites. Known as "camp robbers," they sometimes filch food right from an unwary hiker's hand.

Like jays, crows, and other corvids, nutcrackers are extremely bright. Their intelligence no doubt contributes to their opportunistic streak, an advantage for eking out a living at timberline. Nutcrackers primarily consume conifer seeds, from whitebark pine, ponderosa pine, larch, and juniper. A nutcracker will cram up to 90 seeds in a special pouch under its tongue, like a chipmunk, while holding several more in its long, pointed bill.

Nutcrackers are as industrious as ants, stashing seed caches from midsummer through fall. Up to 33,000 seeds may be stored in a single cache. Studies have determined that the birds remember the location of their stores by taking bearings on rocks and other landmarks. Nutcrackers must recover at least one-third of their seeds to survive a Sierra winter, but plenty remain buried, and that's a good thing. Whitebark pine in particular relies on nutcrackers to bury the seeds in the precise locations and at the precise depths that are ideal for germination.

The American dipper (above) is equipped with a special membrane that covers its eyes and nostrils when it forages underwater.

Alpine ponds (opposite, above) and creeks are good places to look for western tanagers, warblers, flycatchers, and other breeding songbirds in late spring and early summer.

Clark's nutcrackers (below) issue several calls, some almost mellifluous. Another is a squawking *kar-r-ack*.

Old-Growth Wilderness

in his crimson, sunshine-yellow, and glossy-black plumage. The female is also striking, with plumage that runs from greenish-yellow to near-chartreuse. Drawn to the thick canopy, tanagers are usually heard first, then seen; their song, a staccato series of notes, is similar to, but hoarser than, the robin's. Tanagers are territorial, and the males sometimes chivy one another through the foliage. Other birds that favor riverine woodlands and surrounding forests include the Bullock's oriole and such warblers as the MacGillivray's and yellow-rumped. In the higher river canyons and surrounding granite shield lands, Clark's nutcrackers, handsome birds related to crows and jays, forage for pine seeds.

The river canyons, particularly the Tuolumne, are also rattlesnake country. Western rattlesnakes, the dominant species, forage from late spring to early summer. They favor rocky areas and often settle near water to cool off during the heat of the day. Western rattlesnakes are fascinating and beautiful animals, but, like all venomous reptiles, their "image" suffers from bad press. The truth is they are extremely diffident and will never threaten human beings unless cornered or surprised.

Away from the rivers, the park is a mosaic of rolling granite, open meadows, thick coniferous forest, and oak groves. Topography determines all – distinctive vegetation zones are found at different elevations. At 3,000 to 7,000 feet, pine and fir forests dominate, harboring a fascinating array of specialized animals. **Badger Pass**, 20 miles south of **Yosemite Valley**, and **Crane Flat**, about six miles southeast of the **Big Oak Flat Entrance** to the park, have significant stands of old-growth trees – the favorite haunt of the great gray owl, one of California's rarest birds and, at 27 inches long, the largest owl in North America. About 80 percent of the state's breeding population of great grays are found in the park. These huge relatives of the endangered

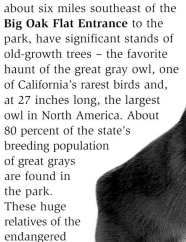

Coyotes hunt small mammals in meadows, rising up on their hind legs to pounce on their quarry.

spotted owl are noted for their spherical heads, broad, disk-shaped faces, and penetrating yellow eyes. Often hunting by day, as well as at dawn and dusk, they pounce on rodents from their perches in deep conifers. Your chances of sighting them are enhanced if you move slowly through heavy old-growth timber, stopping often to scan lower limbs with binoculars. Northern goshawks – fierce, rare accipiters of the deep woodlands – share the great gray owl's habitat and are sometimes spotted darting through the timber.

Fisher and marten, two rare forest mustelids, also live in stands of old-growth conifers, and the fortunate (and exceedingly observant) visitor may sight one or the other, or at least their tracks, at **Badger Pass** or **Crane Flat** or **Wawona**, at the south end of the park. The fisher, a large, sinuous dark-furred animal, is one of the few predators adept at killing porcupines. It flips them over and attacks the underbelly,

Striped skunks (left) are nocturnal feeders with a broad diet, from various insects and carrion to eggs in an unattended nest.

Mountain lions (below) are rarely seen; visitors are more likely to discover their tracks.

where quills are nonexistent. The smaller, more arboreal, yellowish-brown marten subsists primarily on squirrels. Both animals are more active in the day than most other predators. To maximize the chances of a sighting, stealthily explore groves of large timber at dawn and dusk. Look for their tracks along creeks or at the bases of trees. Fisher tracks have five clawed toes and a somewhat shortened, irregularly shaped rear pad. Marten tracks display five dainty clawed toes and a small, crescent-shaped pad imbedded in an oval depression caused by their lush fur.

Fisher and marten, as well as other species that thrive on timber habitat, also find a haven in one of the state's last stands of old-growth red fir, in an area east of **Crane Flat** alongside **Tioga Road**. The trunks of some of these beautiful russet-barked conifers approach six feet in diameter. Although the terrain is steep, hiking among these stately giants is relatively easy, due to the openness of the understory. The cathedral-like quality of these groves is reminiscent of the giant sequoia forests along the King's Canyon drainage farther to the south.

Porcupines, the fisher's favored prey, make for fascinating observation in their own right. Yosemite supports a good population of these hefty quilled rodents. They waddle along roads and trails, and perch in pine trees, gnawing away the tough outer bark to get at the tender, delectable (if you're a porcupine) cambium. Their scat is highly distinctive: a big, brown pellet composed of compressed pine bark.

Yosemite's backcountry is the territory of one of the rarest mammals in America – the Sierra Nevada bighorn sheep, a race of the California bighorn. Unfortunately, this wild ovid is susceptible to livestock diseases, and predation by increasing numbers of mountain lions has taken a devastating toll on lambs and young ewes. There are now fewer than 100 sheep left in the entire cordillera, no more than 20 or 22 in Yosemite. The lucky visitor hiking around **Tioga Crest** on the **Saddlebag Lake Trail** north of **Tioga Road** will catch a glimpse of these reclusive animals warily watching from a distance.

While not as charismatic as Sierra Nevada bighorns, rare and noteworthy amphibians also occupy Yosemite's hinterlands. The Yosemite toad is a resident of the true High Sierra, preferring mountain meadows and margins of alpine woodlands. Toads emerge shortly after the snow melts and breed in shallow lakes and tarns. Although these relatives of the western toad have experienced a mysterious and alarming decline, their trilling antiphonies still punctuate the twilight hour. The mountain yellow-legged frog is a denizen of high mountain streams and small pools charged by running water. These handsome, if not particularly showy, animals have dark-spotted ocher or rufous dorsal surfaces and bellies and underthighs that range from lemon yellow to orange. When touched, they emit a distinct, rather pleasant garlic odor.

Most visitors to Yosemite stay close to their cars, content to admire the spectacular scenery of the park's legendary landmarks. But a short hike can leave the crowds behind, revealing the immutable, pristine Yosemite of John Muir.

Martens (above) are diurnal predators. They dwell in trees and subsist primarily on squirrels.

Pikas (left) can be heard bleating from within their dens in the high country, and yellow-bellied marmots can be seen sunning themselves on rocks.

TRAVEL TIPS

DETAILS

When to Go

Avoid summer crowds by visiting in late spring or early fall. If you must come in summer, reserve lodgings six months to a year in advance. Summer highs in Yosemite Valley sometimes reach 95°F; expect thunderstorms. Winter is relatively mild in the valley, but snowfall is heavy at higher elevations. Weather can change suddenly at any time of year, so be prepared for a wide range of conditions.

How to Get There

Fresno Yosemite International and Merced Municipal are the nearest airports. Yosemite Sightseeing, 209-443-5240, offers bus service from Fresno and Merced. Yosemite Via, 888-727-5287, offers bus service from Merced all year and from Fresno in summer.

Getting Around

Car rentals are available at both airports. Free shuttle bus service is available year-round in Yosemite Valley.

Backcountry Travel

A free backcountry camping permit is required. To obtain one, send a request to Wilderness Permits, Box 545, Yosemite National Park, CA 95389, or call 209-372-0740. Include your name, address, phone number, number of people, method of travel, start and finish dates, and desired trail route.

Handicapped Access

Visitor centers, nature center, and select trails and campsites are accessible.

INFORMATION

Yosemite National Park
P.O. Box 577; Yosemite, CA 95389; tel: 209-372-0200.

California State Division of Tourism
801 K Street, Suite 1600; Sacramento, CA 95814; tel: 800-462-2543 or 916-322-2881.

CAMPING

There are 13 campgrounds in the park and over 1,400 campsites. Some may be reserved in advance, others are first-come, first-served. For reservations, contact the National Park Reservation System, P.O. Box 1600, Cumberland, MD 21502; tel: 800-436-7275.

LODGING

PRICE GUIDE – double occupancy

| $ = up to $49 | $$ = $50-$99 |
| $$$ = $100-$149 | $$$$ = $150+ |

Ahwahnee

Yosemite Concessions Services, 5410 East Home Avenue; Fresno, CA 93727; tel: 209-372-1488 or 209-372-1489 or 209-252-4848 (reservations).

This extraordinary granite-and-log structure has been designated a National Historic Landmark. Ansel Adams described it as "one of the world's distinctive resort hotels." Decorated with American Indian designs, the hotel's rooms have comfortable furnishings and private baths; some have king-sized beds and glorious views of Glacier Point and the Valley's south wall. Amenities include a restaurant, afternoon tea in the Great Lounge, gift shops, pool, and tennis courts. $$$$

Curry Village

Yosemite Concessions Services, 5410 East Home Avenue; Fresno, CA 93727; tel: 209-252-4848.

Situated in the shadow of Glacier Point, Curry Village offers 426 tent cabins, 180 cabins, and eight hotel rooms. Wood-and-canvas cabins sleep up to five and have linens, blankets, and maid service; guests share centrally located bathroom facilities. Freestanding wood cabins, with and without private bath, accommodate up to five people and have maid service. Amenities include a cafeteria, pool, ice-skating rink, gift shops, and bicycle rentals. Open spring to fall and on weekends and holidays in winter. $–$$

Karen's Bed-and-Breakfast

1144 Railroad Avenue, P.O. Box 8; Fish Camp, CA 93623; tel: 800-346-1443 or 209-683-4550.

Just two miles south of the park, this inn offers three guest rooms decorated in contemporary country style. Each room has a private bath and views of the surrounding woods. $$

Yosemite Lodge

Yosemite Concessions Services, 5410 East Home Avenue; Fresno, CA 93727; tel: 209-252-4848 or 209-372-1269.

Affording good views of both Yosemite Falls and the Merced River, this lodge's 495 units offer four types of accommodations: deluxe rooms with balconies or patios, standard hotel rooms with or without a private bath, and rustic cabins with or without a private bath. A restaurant, a lounge, a pool, gift shops, and bicycle rentals are available. $$–$$$

Yosemite View Lodge

11136 Highway 140; El Portal, CA 95318; tel: 800-321-5261 or 209-379-2681.

Most of the lodge's 278 rooms have kitchenettes, and over 130 have spa tubs for two, fireplaces, and private balconies overlooking the Merced River. Two restaurants, a bar, three pools (one indoor), and a gift shop are also on the premises. $$–$$$$

TOURS & OUTFITTERS

High Sierra Goat Packing

P.O. Box 82; Columbia, CA 95310; tel: 209-536-9576.

Goat-packing and fishing trips in the Emigrant, Carson-Iceberg, and Hoover Wilderness Areas, which border Yosemite National Park. Goats are excellent pack animals and can carry 50 to 90 pounds. They are sure-footed and docile and have little impact on the environment, and they can be taken into areas that other animals are unable to reach. Trips include gourmet meals.

Yosemite Association

P.O. Box 230; El Portal, CA 95318; tel: 209-379-2646.

Naturalists conduct more than 50 seminars each year on such topics as birding, geology, and astronomy. The association also leads two- to seven-day backpacking excursions in the park. Call or write to request a catalog.

Yosemite Concessions Services

Yosemite, CA 95389; tel: 209-372-1240.

Daily guided sightseeing tours in open-air trams and motor coaches. One- to eight-hour tours describe the park's geology, flora, and fauna.

Yosemite Valley Stables

Yosemite, CA 95389; tel: 209-372-8348.

Two-hour, four-hour, and all-day riding excursions on mule or horseback through Yosemite's beautiful backcountry; four- to six-day saddle trips through Yosemite's High Sierra; and open-ended pack trips in the park. Tours can accommodate both novice and experienced riders.

Excursions

San Luis National Wildlife Refuge Complex

P.O. Box 2176; Los Banos, CA 93635; tel: 209-826-3508.

This complex north of Fresno includes the San Luis, Merced, Kesterson, and Joaquin River refuges. Great numbers of shorebirds and waterfowl join bald eagles, sandhill cranes, marsh wrens, endangered Swainson's hawks, and many other bird species. Marsh and grassland sustain a variety of mammals, including tule elk, the smallest subspecies of North American elk, or wapiti.

Sequoia and Kings Canyon National Parks

Three Rivers, CA 93271; tel: 209-565-3134.

This is truly a land of giants – gaping canyons, hulking mountains, powerful rivers, and the world's largest organism, the giant sequoia. This vast wilderness park preserves the rugged part of Sierra Nevada, and is home to all types of mountain creatures, from diminutive pikas to stout black bears. Visitors can enter by car but most of the park is accessible only on foot. The backcountry is laced with hiking trails, including a section of the John Muir Trail, which runs 219 miles from Yosemite Valley to Mount Whitney, the highest peak in the continental United States.

Sierra National Forest

1600 Tollhouse Road; Clovis, CA 93611; tel: 800-280-2267 or 209-297-0706.

Situated on the western slope of the Sierra Nevada between the Merced and Kings Rivers, the forest's 1.3 million acres harbor 315

species of wildlife, including black bears, bobcats, beavers, coyotes, and gray foxes, to name but a few. Five native mule-deer herds graze here, and more than 480 lakes and 1,800 miles of streams and rivers provide habitat for creatures such as the red-legged frog and water shrew. The forest also protects three sensitive birds: the California spotted owl, willow flycatcher, and goshawk.

Olympic
National Park
Washington

CHAPTER **19**

The banana slug is normally not quite the size of a banana, and never a banana's bright, sunshiny yellow. It more resembles a wan, olive-colored cigar that oozes a wake of mucuslike slime as it scoots across a rain forest trail at a top speed of 30 feet per hour. It loves dank, sullen weather, as sunlight interferes with its production of slime, which lubricates its movement and disgusts predators with a bitter flavor. It is notoriously undiscriminating in its choice of a mate. Either sex will do since a banana slug produces both sperm and eggs, and sometimes a couple will fertilize each other in a single all-day spasm of rapture. ◆ All of which is why nature aficionados visiting Washington's scenic **Olympic National Park** should go ahead and wrap themselves in Gore-Tex on the gloomiest day, take a trail into the **Hoh**, **Queets**, or **Quinault** rain forests, and pause for a respectful contemplation of the first banana slug they see (which won't take long). What better example of the astonishing diversity

The quest for wildlife takes travelers deep into rain forests, up mile-high mountains, and along coastal waters traversed by whales.

of life on earth, the endless imagination of Nature to meet her obligations no matter how outrageous the means? A plus for slug watchers: No animal submits to photography more patiently. ◆ Olympic National Park pre-serves 1,400 square miles of mountains and rain forests, and a long, skinny strip of the Pacific Coast of the **Olympic Peninsula** – very, very fortunately. If President Franklin D. Roosevelt hadn't come out for a personal tour, the entire 6,500-square-mile peninsula might today be shorn like a Marine recruit's skull. In 1937, staring through a car window at the desolation of clear-cut forest, FDR fumed, "I hope the son-of-a-bitch who logged that is

Temperate rain forests draped with moss provide habitat for a wide range of creatures, from the majestic bald eagle to the banana slug.

roasting in hell!" The following year, he signed the bill creating the park. "The trees alone are enough to bring humility to man," wrote William O. Douglas, who divided his life between the U.S. Supreme Court and his beloved Northwest cathedrals of Sitka spruce, Douglas fir, and red cedar 200 to 300 feet high. The trees aren't alone. Olympic National Park comprises a world of extravagant ecosystems: tide pools supporting 700 species of plants and animals, rivers glittering with salmon, alpine meadows grazed by deer and elk, craggy peaks and creaky glaciers patrolled by red-tailed hawks and bald and golden eagles. Nature took an audacious gamble here, packing so much into one small space that *Homo sapiens,* as is our custom, would be tempted to overrun the place. But the park is providentially located in a far corner of the continent penetrated by few roads. Humans have made a difference here, but not, so far, a ruinous one.

Wild Creatures at High Altitudes

A safari into the park could begin anywhere, but the single best opportunity, **Hurricane Ridge**, also happens to be the most accessible and offers the most stunning scenery as a backdrop. One caveat: Snow depth in winter averages 10 feet, and 80-mile-per-hour winds sometimes rake the ridge. A hurricane with snow is something to ponder. The 17-mile drive from **Port Angeles** to the visitor center is on a good paved road that winds through layered weather en route to a 5,242-foot elevation. Clouds snooze in the defiles like ghosts lounging in hammocks; fog and drizzle alternate with what Northwest people cheerfully call "sun breaks." On a good day, the view south from the ridge parking lot is among the most glorious sights in Washington, as a meadow falls away from the ridge in two gibbous arcs, a distant curtain of conifers rakes the sky, and a frosted band saw of mountains forms the horizon.

Black-tailed deer, unfortunately, sometimes graze in the parking lot, attracted by the sweet and toxic ethylene glycol burped from overheated radiators. Rangers do their best to clean it up.

Three paved, wheelchair-accessible nature trails loop north from the lot, offering visitors a chance to view the ubiquitous deer close-up on their natural turf. Oblivious to the human parade passing 10 feet away, they forage, stop periodically to stare in the direction of distant sounds, and kick each others' butts, literally, in what appear to be frequent disputes over grazing rights. Several times a week, black bears amble to within 50 or 100 yards of these trails, snacking on tender new subalpine fir tips in the glacial cirque below the ridge. Bears are abundant enough on the peninsula that the park service occasionally closes some backcountry to backpackers, but there's never been a problem at Hurricane Ridge.

The road curls another 1.3 miles to **Hurricane Hill**, which then offers an indispensable 1.5-mile (each way) trek to a spectacular mountain spine above tree line. On a clear day, the **Strait of Juan de Fuca** and **Vancouver Island** are visible to the north; the 10,778-foot volcanic cone of Mount Baker rises nearly 100 miles to the east. This hike, a little more strenuous than the nature trails at the ridge, thins out the human population, and the meadows below are so vast and open that they're a virtual showcase for wildlife. A hiker's binoculars pick out a trio of Roosevelt elk, the bull a monstrous five-point brawler that could weigh as much as 1,000 pounds. Not only the elk's size corresponds to nutrition, so does the male's

Black-tailed deer (opposite) are among the most commonly seen animals at Hurricane Ridge.

Olympic marmots (left), so named because the species is found only in the Olympic Mountains, are seen by hikers who venture into the high country.

headgear – which on this specimen is a living testament to the park's protected forest buffet. The Northwest's largest herd of Roosevelt elk (named for Teddy, not FDR) roams the Olympic Peninsula; other, smaller herds graze Vancouver Island and northern California.

A few steps farther toward the clouds, a huddle of hikers forms to watch an Olympic marmot, another local subspecies that seems to grow larger – up to a foot and a half, not counting tail – than its Cascade relatives. This one seems all but blasé about its audience, digging out a root with such rapid strokes of its forepaws that they almost vibrate, but every minute or so it stops, rises on its hind legs, and stares at the crowd. Like any tasty rodent, it frequently evaluates its surroundings to see whether there is anything that looks like a predator around. If a bald eagle had been circling, the marmot would have shrieked its alarm ("whistling marmots," as they're sometimes called, use their vocal cords, not their teeth) and scampered for its burrow.

Other frequent Hurricane Hill sightings include Douglas squirrel, Olympic chipmunk, snowshoe hare, gray jay, blue grouse, red-tailed hawk, and golden and bald eagles. Mountain lions, also called cougars, are always around, but they're so furtive that one ranger has sighted only one in 16 years in the park. A hundred years back, the gray wolf was a vital part of this ecosystem, but the last one in the Olympics was trapped in 1924. A proposal to import a pack from Vancouver Island is under study, and it has some vigorous allies.

Tide Pooling and Whale Watching

It's a long and twisty 100 miles from Hurricane Ridge to the park's 60-mile-long Pacific Coast strip, but this is the opportunity to visit a dramatically different ecosystem on the wildest ocean shoreline remaining in the continental United States. It's so wild, in fact, that every summer the park service has to pluck unwitting adventurers from rocks and cliffs where they've fallen or been trapped by a high tide. (Check the day's tide table before wandering away from any of the beach access points.)

This coastline induces humility. In winter, gunmetal-colored storms brood and drizzle, then cut and slash, hacking sea stacks into abstract sculptures. Drift logs weighing several tons loll on the beach, waiting to be tossed like pencils in the next storm. An especially adventurous subspecies of human actually likes to walk and even camp on these beaches in midwinter, just to experience Nature expressing her most primal urges. In any season, the Olympic seascape seems more awesome than simply beautiful; so much of it is off the scale of conventional human experience. But spying on tide pools yields a wealth of intimate pleasure. They abound in colorful life, which park naturalists point out during intertidal beach walks offered daily in summer.

On one of these walks, a naturalist plucks a whorled triton's shell from a pool and places it in a small boy's upturned hand. As if on cue, a hermit crab, looking like a bleached, dime-sized spider, stumbles into the sunlight, takes one look at the monster holding it, then shrinks back into its pirated shell. Moments pass. The crab reconsiders – if this is something an arthropod with a brain no larger than a pinhead can be said to do – and emerges to explore this strange pink world, carrying its shell like a backpack. The naturalist returns the crab to the water (taking any living creature from a park tide pool is illegal) and points out the prolific sea life in its bathtub-sized neighborhood. Hordes of glossy blue mussels cling hopefully to pitted rocks exposed by the receding tide. Giant green anemones, enchanting in their almost translucent lime

Bat stars (opposite) remain anchored in rough surf by attaching their minute suction cups to rocks.

Pacific tree frogs (right) climb trees with the aid of their large toe pads but often stay close to the ground, near water.

hue, seem to our intuition ominous and malevolent. What creature waving 300 tentacles wouldn't be? Clusters of vermilion sea stars cling to the rocks, many with their points wrapped in another's embrace.

Beach 4, about four miles north of the **Kalaloch Information Station**, provides the best tide pooling on the coastal strip. **Rialto Beach**, **Ruby Beach**, and **Hole-in-the-Wall** also leave good pools at low tide. But any of the park's beaches offer opportunities to spot more mobile creatures. Twenty-nine species of marine mammals live in or pass through the Olympic coastal waters. Sea lions and harbor seals claim the offshore rocks; sea otters, once hunted to extinction off this coast, began a slow comeback in 1969–70 when 59 new animals were imported from Alaska.

The prize catches for sightseers, of course, are the gray whales, which commute past the Olympic coastline en route to the Bering Sea from March through May, then back to Baja in October and November. Traveling in pods of two to twelve, they cruise at about four knots on their 6,000-mile (each way) migration, the longest undertaken by any mammal. The spring parade offers better whale watching, because the whales, for some reason, swim closer to shore. Your

The Ancient Forest

The ancient forests of the Northwest are almost gone, reconfigured into flimsy houses for a booming world population. A 1989 study estimated that only 16 percent of the virgin wood on the Olympic Peninsula, America's once-heroic forest, remained. Second-growth forest and "managed" timberland lack the astonishing fecundity and diversity of old growth. Missing, among many things, are "nurse" logs, fallen trunks encrusted with moss, ferns, and seedlings of future skyscraper conifers.

There is more here than aesthetic loss to humans who like to wander the forest trails. Old growth is an irreplaceable wildlife habitat. Those rotting logs, for example, provide a haven for carpenter ants and bark beetles, which in turn nourish woodpeckers. Sever one link of a complex ecosystem, and the ripples resonate throughout the entire web – sometimes forever.

The most notorious ripple in the Northwest forest is the northern spotted owl, a reclusive, nocturnal raptor with a four-foot wingspread and a strident, almost doglike hoot. It landed on the endangered list in 1973, the same year Congress approved the Endangered Species Act. A flurry of studies over the next decade reached a watershed conclusion: the owl was threatened because old-growth forest was threatened. It needs old growth to make its living, perching on snags and swooping between big, widely spaced trees for its prey. The shy bird grew into a symbol of the conflict between wilderness preservation and economic growth. In 1992, a federal court halted logging of the spotted owl's habitat, which shook the timber-dependent economy of the peninsula to its heartwood.

"In wildness is the preservation of the world," Thoreau wrote. But on the uncrowded continent of 1862, that was both easier to say and less important to do than it is today.

Backcountry camping (opposite) brings visitors close to wildlife and, in some places, offers incomparable views of snow-capped peaks.

Spotted owls (below), because of their retiring nature and the density of old-growth forests, are difficult to find but can be heard by determined and patient observers.

best chance is to select a calm, gray day and locate a tall headland jutting into the sea, such as **Point of the Arches** or **Shi Shi Beach**. Scan the water up to a mile off shore, and bring patience, as well as binocs.

The Banana Slug's Forest

The Olympic rain forests are places of deep and melancholy silence; a single creature's rustle or whistle sometimes can seem like an event. A pioneer lamented in his journal, "It is seven months since I recall a songbird." But vigilant visitors will hear, and see, plenty of wild creatures. Elk herds migrate down from the mountains in winter and spring, and the predators – mountain lions, bobcats, bears – are always about. The forest floor is alive with restless shrews (they never sleep) and Douglas squirrels (they never hibernate). The trees are home to pileated woodpeckers, Steller's jays, winter wrens, nuthatches, and many more species. Of course, the rain forest would be replete with small animal life; the same dense foliage that makes sightings tough for wildlife watchers also frustrates predators.

The short, crowded **Hall of Mosses** trail at the **Hoh Rain Forest Visitor Center** provides a quick rain-forest immersion. Better opportunities to see wildlife – and enjoy much quieter, contemplative walking – can be found on the **Hoh River Trail** and **Sams River Loop Trail** at the end of the **Queets River Road**.

The Olympic rain forests begin just a few miles inland, so it's possible, and a fine lesson in ecology, to observe both gray whales and banana slugs within a few minutes of each other. Each has mastered what it must do, and therefore each is equally noble. Life on Earth is almost incomprehensibly wonderful and weird, and there may be no better place than the Olympic Peninsula to contemplate this.

TRAVEL TIPS

DETAILS

When to Go

Summer is the driest season, but be prepared for rain year-round, especially in the wet, western half of the park. Summer highs range from 65° to 75°F, lows from 45° to 55F°. Winter is mild, with temperatures at lower elevations in the 30s and 40s. Snowfall is heavy and temperatures far cooler at high elevations.

How to Get There

Seattle-Tacoma International Airport is a three-hour drive from the park. Fairchild International Airport is located in Port Angeles, near the northern boundary of the park. Greyhound Bus Lines, 800-231-2222, and Olympic Bus Lines, 360-452-3858, operate between Seattle and Port Angeles.

Getting Around

Car rentals are available at both airports. The park has several scenic roads, though none penetrates the interior. There are, however, 600 miles of hiking trails. Tours are provided by Olympic Van Tours, 360-452-3858.

Backcountry Travel

A wilderness use permit is required for all backcountry camping; reservations are required in some areas. Permits may be purchased at ranger stations or the Wilderness Information Center in Port Angeles. For reservations and information, call 360-452-0300.

Handicapped Access

Visitor centers at Port Angeles and Hoh Rainforest are accessible, as are several nature trails and select campsites (some assistance may be required). Wheelchair access is limited at other ranger stations and visitor centers.

INFORMATION

Olympic National Park

600 East Park Avenue; Port Angeles, WA 98362; tel: 360-452-4501.

North Olympic Peninsula Visitor and Convention Bureau

P.O. Box 670; Port Angeles, WA 98362; tel: 800-942-4042.

Washington State Tourism

P.O. Box 42500; Olympia, WA 98504-2500; tel: 800-544-1800 or 360-586-2088.

CAMPING

There are 16 campgrounds in the park. Most facilities provide water, toilets, and garbage containers. Individual campsites offer a picnic table and fire pit or grill. Campsites are available on a first-come, first-served basis. Call 360-452-0330 for information.

LODGING

PRICE GUIDE – double occupancy

$ = up to $49	$$ = $50-$99
$$$ = $100-$149	$$$$ = $150+

Kalaloch Lodge

157151 Highway 101; Forks, WA 98331; tel: 360-962-2271.

Perched on a seaside bluff within Olympic Coast National Marine Sanctuary, this rustic lodge offers 58 units, including 18 ocean-view cabins and a secluded 10-room motel. The main lodge has comfortable rooms and a few suites. The cabins have kitchenettes and fireplaces and can sleep four to nine people in basic comfort. A restaurant, lounge, coffee shop, and gift shop are also on the premises. $$–$$$

Lake Crescent Lodge

416 Lake Crescent Road; Port Angeles, WA 98363; tel: 360-928-3211.

This two-story cedar-shingle lodge, built in 1916, sits at the foot of Mount Storm King, just steps away from Lake Crescent. Accommodations are simple but comfortable. Five rooms in the old lodge have shared baths. Thirty motel rooms and 17 modern cabins have private baths. All cabins have decks; four have fireplaces. Amenities include a lakeside restaurant, a cocktail lounge, rowboat rentals, and a gift shop. $$–$$$

Lake Quinault Lodge

P.O. Box 7; Quinault, WA 98575-0007; tel: 800-562-6672.

Franklin Delano Roosevelt once stayed at this cedar-shake lodge, built in 1926 on Lake Quinault in the Hoh Rain Forest. The resort's 92 rooms are tastefully appointed with private baths, some with fireplaces. Exposed beams of Douglas fir, an enormous brick fireplace, and antique wicker furniture give the lobby a grand rustic feeling. Amenities include a restaurant, a gift shop, boat rentals, an indoor heated pool, and saunas. $$–$$$$

Lizzie's

731 Pierce; Port Townsend, WA 98368; tel: 800-700-4168.

Built in 1887, this Victorian bed-and-breakfast has seven guest rooms. Various touches include an Art Nouveau bed, etched-glass French doors, an Eastlake queen-sized bed, plaster ceiling friezes, and a bay window. Two grand parlors, each with fireplace, are furnished with leather sofas. Children enjoy a Victorian playhouse (a replica of the main house) and a secret garden. Five rooms have private baths, four with claw-foot tubs. A 10-minute walk from historic downtown. $$–$$$.

Sol Duc Hot Springs

P.O. Box 2169; Port Angeles, WA 98362-0283; tel: 360-327-3583.

Soak in the sulfur hot springs at this secluded resort, built in 1910 in the northern section of the park. The 32-unit complex includes cabins with full bath and two double beds; some have kitchens. A campground with sites for recreational vehicles is available. Amenities include a poolside restaurant, bar, deli, massage therapy, gift shop, and grocery store. $$

TOURS & OUTFITTERS

Olympak Llama Tours

1614 Dan Kelly Road; Port Angeles, WA 98363; tel: 360-452-5867.

Completely outfitted llama pack trips of a day or longer.

Olympic Outdoor Center

18971 Front Street; Poulsbo, WA 98370; tel: 360-697-6095.

Sea kayaking off the Olympic Peninsula, with guided trips and instruction available.

Olympic Raft and Guide Service

239521 Highway 101 West; Port Angeles, WA 98363; tel: 360-452-1443.

Guided float trips on the Elwha and Hoh Rivers, as well as sea and lake kayaking.

Olympic Wilderness Tours

5196 Lars Hansen Road Southeast; Port Orchard, WA 98367; tel: 360-871-9087.

Guided and outfitted backpack and mountain-climbing trips anywhere within the park.

Quinault Packing

42 Liscumm Road; Quinault, WA 98575; tel: 360-288-2240.

Custom-designed donkey pack trips of any length.

Whale Watching Adventure

431 Water Street; Port Townsend, WA 98368; tel: 360-385-5288.

All-day whale-watching tours of the San Juan Islands, with frequent sightings of killer whales, porpoises, and eagles.

Wooley Packer Llama Co.

5763 Upper Hoh Road; Forks, WA 98331; tel: 360-374-9288.

Customized llama packing trips in the park, overnight to seven days.

Excursions

Mount Rainier National Park

Tahoma Woods, Star Route; Ashford, WA 98304; tel: 360-569-2211.

Glaciers creep down the slopes of Mount Rainier, the tallest peak in the volcanic Cascade Range. Ancient forests, lush wildflower meadows, and cold, rushing rivers sustain nearly a thousand species of plants and animals. Large mammals include mountain goats, Roosevelt elk, black-tailed deer, and black bears. Only the luckiest visitors will spy stealthy hunters like lynx and mountain lions.

Nisqually National Wildlife Refuge

100 Brown Farm Road; Olympia, WA 98516; tel: 360-753-9467.

Situated on a 4,000-acre river delta on Puget Sound, Nisqually is a vital stopover for thousands of migrating shorebirds, waterfowl, and raptors. The refuge welcomes grebes, scoters, goldeneyes, green-winged teals, American wigeons and dunlins, and other species of dabbling and diving ducks, not to mention bald eagles and great horned owls. Also active here are harbor seals, river otters, coyotes, beavers, minks, and cottontail rabbits.

North Cascades National Park

2105 Highway 20; Sedro Woolley, WA 98284; tel: 206-856-5700.

Fury. Despair. Terror. Challenger. These are just a few of the names given to the glacier-carved peaks in this remote and rugged corner of the Cascade Range. Wildlife abounds. Black-tailed deer graze on grassy mountain slopes; black bears feed on berries in subalpine meadows; eagles soar on air currents above sawtooth peaks; mountain goats clamber up narrow ledges; and mountain lions prowl the high country. A single road traverses the park, but nearly 400 miles of hiking trails lead into the backcountry.

Kenai Peninsula
Alaska

For several hours following their drop-off from a charter boat, a couple of sea kayakers paddle quietly up **Aialik Bay** in **Kenai Fjords National Park**, a vast teal-blue expanse of water studded with icebergs. In the first hour, a 60-foot humpback whale rises from the surface in a barnacled mass, and a pod of Dall porpoises swims nearby. Several sea otters roll in the water, grooming, and a black bear digs for roots in an avalanche slide. As the morning mists clear, the kayakers glimpse for the first time the snowy peaks that tower 3,000 feet above the bay, the slopes struck here and there by shafts of brilliant sunlight.　◆　In the early afternoon, the kayakers come within sight of glistening **Aialik Glacier**, a fortresslike structure of translucent blue ice rising more than 700 feet above the water. When a chunk occasionally breaks off, a massive swell sweeps down the bay. After the kayakers rejoin their charter boat, they head to the waters at the mouth of the bay. There, on a small rocky island, black-legged kittiwakes,

Exploring the Kenai reveals puffins nesting on sea stacks, sea otters diving in fjords, and brown bears poised on tundra ridges.

horned puffins, and glaucous-winged gulls stand vigilantly over their nests. A little farther on, more than a thousand seals occupy another island, for this is early June, the peak of the seal-pupping season. As the boat slips away, watched by the baby seals and their mothers, the passengers wonder if this spectacle of beauty and abundance was how the world had once been.　◆　From fjords like Aialik in the south to the lakes and wetlands in the north and along the many mountains and valleys, **Kenai Peninsula** is a place of wildlife superlatives: an estimated three million migratory birds pass through each year, salmon runs fill local rivers, Yukon moose and

Horned puffins, seabirds that nest on rocky islands, gather fish in their large, colorful bills as they "fly" underwater. Tufted puffins, a related species, are also found in the peninsula's waters.

the world's largest brown bears inhabit the mountains, and half a million seabirds of 30 species occupy the waters and islands off the coasts. Because of the peninsula's location at the edge of the **Gulf of Alaska**, the surrounding waters are particularly attractive to marine life, as the nutrient-rich ocean currents upwell from the continental shelf. As a result, vast numbers of seabirds and marine mammals gather in the bays and sea islands of the Kenai. Islands, cliffs, and sea stacks are host to colonial nesting seabirds such as puffins, auklets, petrels, murres, and kittiwakes. Steller's sea lions haul out on rock benches at tide line. Harbor seals

doze on drifting icebergs. Dall and harbor porpoises, sea otters, and humpback, minke, and orca whales busily work the waters of the bays and fjords. Salmon literally fill coastal streams, rivers, and lakes as they return to spawning waters, and in their plenitude attract everything from brown bears to bald eagles.

Alaska Wildlife Sampler

This peninsula within easy reach of **Anchorage** extends southward from the coast of central Alaska for more than 130 miles and at its widest spans 110 miles. Cut off from the frozen north by the Alaska Range,

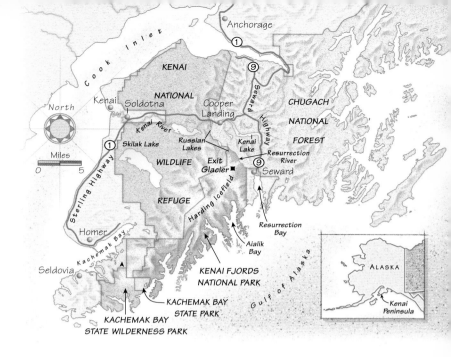

warmed by the waters of the Japanese Current, it is far more temperate than its latitude would suggest. First seen by Captain Cook in 1778 and long occupied by Native Americans before that, the land has never been significantly settled. To this day, there are only a handful of frost-heaved roads. Residents like to cite the fact that there are more tufted puffins on one group of islands (the Channel Islands) than there are people on the entire 12,000 square miles of the peninsula.

Visitors to Alaska often choose to spend their entire trip on the Kenai Peninsula because it offers a microcosm of the state, ranging from a 965-square-mile icefield to vast coastal rain forests, from tundra-covered peaks to glaciated fjords. Additionally, there are millions of acres of public lands: **Kenai Fjords National Park** occupies much of the eastern coast along the Gulf of Alaska. **Kenai National Wildlife Refuge** comprises two million acres from **Cook Inlet** in the north to the mountains and glaciers on the south side of **Kachemak Bay**. The **Chugach National Forest** protects much of the northeastern portion. **Kachemak Bay State Park** near **Homer** is located in the wild southern region of the peninsula.

Most people visit the Kenai during the summer months, when wildlife are most plentiful and the weather is most conducive to wildlife viewing. Large, comfortable wildlife tourboats run daily out of **Seward** into the nearby Kenai Fjords National Park. Arrangements can also be made for single-day or multiday sea-kayak drop-offs. These popular trips range from whale migration routes to puffin colonies to Steller's sea lion haul-outs. In the interior of the Kenai Peninsula, such animals as moose, Dall sheep, and bald eagles are often seen beside

the road, and spawning salmon are easily viewed from the platforms on Portage Creek in Chugach National Forest.

Hiking Amid the Giants

One of the most accessible areas on the peninsula for wildlife viewing is **Exit Glacier**, reached by the access road on the west side of the Seward Highway, four miles north of **Seward**. Here, visitors can walk to the face of the glacier or take a more strenuous four-mile hike up to the immense **Harding Icefield**. The road also leads to the **Resurrection River Trail**. Heading up the trail, which parallels the Resurrection River, hikers almost immediately pass into another world. Precipitation on this east side of the peninsula exceeds 200

Orcas (left), also known as killer whales, travel in pods, find prey using echolocation, and hunt cooperatively.

The short-tail weasel (right) turns white in winter. Its summer coat is dark brown.

John Muir (right), who had seen glaciers in the Sierra Nevada, was among the earliest naturalists to study Alaska's glaciers.

Sockeye salmon (below), cramming streams during their spawning run, are magnets for bald eagles and brown and black bears.

Kayaking (opposite) the peninsula's bays and coves allows visitors to closely observe the habits of otters, seals, and seabirds.

Muir in Alaska

It was naturalist John Muir, following his 1879 and 1880 trips to Alaska, who first told the world, in a series of essays in *Century* magazine, about the wonders of Glacier Bay. Muir, who had already seen many remarkable spectacles of nature, was struck by the beauty of the deep fjords and fronting glaciers. Muir Inlet and Muir Glacier, later named for him, are seen by today's visitors to Glacier Bay National Park.

Muir had never viewed anything like the rich wildlife of southern Alaska. He wrote of "breaching whales" and "porpoises blowing and plunging," of "wild goats fifteen hundred feet up a steep grassy pasture" and rivers choked with "tens of thousands" of salmon. The whole was a spectacle of profusion such that "nothing that I could write might possibly give anything like a fair conception of the extravagance of [wildlife] numbers." "To the lover of pure wildness," he observed, "Alaska is one of the most wonderful countries in the world."

inches per year, and the very scale of the old-grown rain forest fills one with awe. There are western hemlocks and Sitka spruces seven feet in diameter – trees that, sliced through, could make a dining-room table, trees that rise to the height of a 15-story building. Tree-sized saplings grow from the side of wind-felled giants. It is in such protected groves that marbled murrelets nest.

The trail offers excellent birding opportunities – great horned owl, goshawk, spruce grouse, Townsend's warbler, golden-crowned kinglet – and also passes through the habitat of black and brown bears, moose, porcupines, red squirrels, beavers, and river otters. Of particular importance to much of the wildlife in this area are the plentiful salmonberry patches – the Kenai is renowned for its berries. Just over 16 miles beyond Exit Glacier, the trail leaves timberline, where it meets **Russian Lakes Trail** and enters the high country. Hikers can spot Dall sheep, mountain goats, caribou (a small transplanted herd inhabits the area), and perhaps even wolves near lovely Upper Russian Lake. In a state not often known for its developed trails (picture Massachusetts in 1790, or Montana in 1890), the Resurrection River Trail into the heart of the Kenai Peninsula is one of the best.

Floating the Kenai

The **Kenai River**, north of the Resurrection River, flows west from powder-blue **Kenai Lake**, which is born in the peninsula's rugged glacial mountains and whose striking color is a result of the glacial till. The river runs fairly straight and level for 50 miles, passing through only one other lake before emptying into the sea at Cook Inlet. A day-trip by raft down the Kenai River offers superb wildlife viewing, especially early in the day, when animals are more active. Most begin their journey at **Cooper Landing**, just below Kenai Lake, and then float the Kenai River 20 miles to **Skilak Lake**.

The river runs fast, between five and six miles per hour, and so the journey is taken at a rapid, keep-your-eyes-open pace. If the trip is made in late August, after the peak salmon season, few other boats are on the river. Dead salmon will already have gathered in the shallows, and black bears, brown bears, and numerous bald eagles appear along the banks to feed on the salmon. Up higher, on the steep tundra ridges, Dall sheep and mountain goats are visible as white dots, and occasionally a brown bear can be seen, appearing in the distance like a golden honeybee. Moose are always plentiful along the Kenai, particularly cows and calves, with bull moose showing up more frequently in May and June. Also during these months, migratory birds such as Canada geese and sandhill cranes pass overhead. Yellowlegs, common and red-breasted mergansers, harlequin ducks, belted kingfishers, and American dippers breed, nest, and raise their young along the river in the late spring and summer.

The outlet of the river at the town of **Kenai** is one of the best areas in Alaska for viewing beluga whales. From spring into fall, pods of the white whales enter the outlet an hour before or after high tide to feed and are visible from a lookout on the south side of town as they move into the mouth of the river.

Adventures in a Roadless Park

Beautiful 171,000-acre **Kachemak Bay State Park**, 50 miles down the coast near **Homer**, is not large by Alaskan standards, but the wildlife is as diverse as the surrounding terrain – steep fjords, gentle wooded hillsides, extensive mudflats. Ferries, water taxis, charter boats, and planes make the roadless

Black bears (right) gravitate to the shores of Kachemak Bay to feed on spawned-out salmon in mid- to late summer.

Harbor seals (below) haul out along the shore and on icebergs.

Caribou (opposite), like this male with a three-foot antler spread, inhabit the peninsula's mountains along with Dall sheep, mountain goats, moose, and bear.

park accessible for day trips. Many people take the ferry over to **Seldovia**, an old settlement on the south shore of the bay, and begin their explorations from there. Others arrange to make the short trip by water taxi or charter boat from Homer or take a naturalist-led trip. Over 80 miles of hiking trails traverse the forest, and the bay is subject to some of the planet's largest tidal fluctuations – 15 to 28 feet – creating many opportunities for observing life in the intertidal zone. Sea kayakers find numerous coves and bays to explore at their leisure. Those who want even more of a wilderness experience, perhaps in the adjoining 200,000-acre **Kachemak Bay State Wilderness Park**, can hop on a float plane in Homer for an aerial trip to one of the backcountry lakes.

One of the commonly seen large mammals in Kachemak Bay State Park is the black bear. At **Humpy Creek** in the northeast area of the park and at other stream fronts, bears fish for salmon. When not scavenging the tidal flats, they browse subalpine areas for blueberries. Trails such as **Sadie Knob** and **Emerald Lake** lead

hikers to the alpine area to see mountain goats and soaring bald eagles and red-tailed hawks. The best places to view moose are the north side of the Kachemak Bay and along **Seldovia Bay**, outside the park, just south of Seldovia. As you will soon notice, everything about the animal is adapted for life in the far north. The long legs enable the moose to travel freely in the deep snow, wade through ponds and marshes, step over windfalls, and browse on high tree limbs. The thick coat protects the moose from the flies and mosquitoes of summer and the freezing cold of winter. Black and massive, the moose browse for their 30 to 50 pounds of edible vegetation per day.

From late spring into summer, seabirds nest in the area and feed in the waters of the bay – 12,000 birds alone on Gull Island, including tufted puffins, common murres, black-legged kittiwakes, and red-faced cormorants. Visitors traversing the bay by boat or scanning the water from shore can spot an array of other birds, from loons to Arctic terns to marbled murrelets, not to mention orca and humpback whales, sea lions, and sea otters. Harbor porpoises slip in and out of the coves and small bays.

The Kenai is a place where the wildlife population is, for lack of a better word, biblical: 10,000 birds on this island, a dozen brown bears on that stream, a myriad of spawning salmon on another river. Caribou and Dall sheep, moose, and mountain goats exist in their natural abundance, wandering as freely as they did when the Russian ambassador first approached President Abraham Lincoln with an offer to sell Alaska. You might glimpse the secretive Canadian lynx or hear the mournful music of the wolf. You can stare eye to eye with a 2,400-pound sea lion or paddle beside a whale that has just risen from the bottom of the sea.

TRAVEL TIPS

DETAILS

When to Go

The vast majority of travelers visit between June and August, when temperatures average in the mid-50s and 60s. Always be prepared for rain and cooler temperatures, however, especially if you plan on exploring the coast.

How to Get There

Major airlines serve Anchorage, about 130 miles north of Kenai Fjords National Park. Year-round bus and plane transportation, and summer rail service, are available between Anchorage and Seward.

Getting Around

Car rentals are available at the airport. Charter flights and cruises can be arranged in Anchorage, Seward, and Homer.

INFORMATION

Alaska Tourism

P.O. Box 110801; Juneau, AK 99811-0801; tel: 907-465-2010.

Anchorage Convention and Visitors Bureau

524 West Fourth Avenue; Anchorage, AK 99501; tel: 907-276-4118.

Kenai Fjords National Park

P.O. Box 1727; Seward, AK 99664; tel: 907-224-3175.

Kenai Visitors and Convention Bureau

11471 Kenai Spur Highway; Kenai, AK 99611; tel: 907-283-1991.

CAMPING

Primitive campsites are located near Exit Glacier and available on a first-come, first-served basis. Backcountry public-use cabins at Holgate Arm, Aialik Bay, North Arm, and Exit Glacier are available by permit only. Cabin stays are limited to three days. To make reservations or request information, call the national park at 907-283-1991. For information about campgrounds outside the park, call the Seward Ranger District at 907-224-3374.

LODGING

PRICE GUIDE – double occupancy

$ = up to $49 $$ = $50-$99

$$$ = $100-$149 $$$$ = $150+

The Driftwood Inn

135 West Bunnell Avenue; Homer, AK 99603; tel: 800-478-8019 or 907-235-8019.

Situated one block from Bishop's Beach, this inn has 21 varied rooms, all clean and cozy, some with shared bath, many with views of the ocean, mountains, and glaciers. A stone fireplace warms the sitting room. Free transportation from airport and ferry. $$

Gwin's Lodge

Sterling Highway; HC64 Box 50; Cooper Landing, AK 99572; tel: 907-595-1266.

The Kenai River is visible from the porches of these rustic cabins and bunkhouses. Each of the lodge's six log cabins contains two queen-sized beds and a private bath. Two simple bunkhouses shelter backpackers and fishermen. Situated 45 minutes from Kenai Fjords National Park and half a mile from Russian River fisheries. $–$$.

Kachemak Bay Wilderness Lodge

P.O. Box 956; Homer, AK 99603; tel: 907-235-8910.

Five-day visits are required at this popular lodge on Kachemak Bay, where private cabins offer modern comfort amid towering spruce trees. The cabins are cozy; each has a tile-and-cedar bath, fine art, a homemade quilt, and a picture window. Gourmet meals, part of the package, are served in a handsome log lodge. Amenities include guide service, hiking trails, a sauna, an outdoor hot tub, and a large dock. $$$$

Kenai Backcountry Lodge

P.O. Box 389; Girdwood, AK 99587; tel: 800-334-8730.

The lodge is set on Skilak Lake in Kenai National Wildlife Refuge and may be reached only by boat. Simple comforts and amenities include a handcrafted log sauna and heated bathhouse. Guests sleep in log cabins or traditional Yukon tents with sturdy roofs, sliding glass doors, front porches, paneled interiors, and heaters. Guided hikes and float trips are also available. $$$$

Seward Windsong Lodge

Mile 6, Exit Glacier Road; P.O. Box 221011; Anchorage, AK 99522; tel: 800-208-0200 or 907-245-0200.

Built in 1996 near the Resurrection River three miles from Seward, this log-sided lodge has more than four dozen rooms, each with two queen-sized beds, natural log furniture, and a full bath. Extras include a restaurant, a bar, and free transportation within the Seward area. $$–$$$

TOURS & OUTFITTERS

Alaska Wildland Adventures

P.O. Box 389; Girdwood, AK 99587; tel: 800-334-8730.

Natural-history tours range from five to 12 days, with rafts, kayaks, and guided hikes. Lodging is provided.

Renown Charters and Tours

507 East Street, Suite 201; Anchorage, AK 99501; tel: 800-655-3806 or 907-272-1961.

Narrated cruises depart from Anchorage and explore Kenai

Fjords or Prince William Sound. Various packages include lunch and visits to the Alaska SeaLife and Native Heritage Centers. The two- to six-hour cruises provide opportunities to view whales (orca, gray, and humpback), Steller's sea lions, harbor seals, bald eagles, mountain goats, puffins, porpoises, and other creatures. Also available are motorcoach charters to various locations and motorcoach, cruise, and rail packages.

Weigner's Backcountry Guiding
P.O. Box 709; Sterling, AK 99672; tel: 907-262-7840.

Fully outfitted day and overnight canoe trips on the lakes and waterways of the Kenai wilderness. Also offers backpacking trips.

MUSEUMS

Alaska SeaLife Center
P.O. Box 1329; Seward, AK 99664; tel: 800-224-2525 or 907-224-3080.

Surrounded by majestic mountains and the splendor of Resurrection Bay, the $52 million, 115,000-square-foot facility is like an open window to the sea. The center, which serves as the gateway to Kenai Fjords National Park, re-creates the habitats of Alaska's marine life in three two-story tanks. Imaginative exhibits explore the habitats of marine birds, Steller's sea lions, fish, harbor seals, and other marine life. Viewing scopes allow visitors to observe a number of marine species, including humpback whales in late spring and early summer.

Excursions

Alaska Chilkat Bald Eagle Preserve

Haines Visitor Information Center; P.O. Box 530; Haines, AK 99827; tel: 800-458-3578 or 907-766-2234.

The preserve was established in 1982 to protect the world's largest congregation of bald eagles. Thousands gather here in fall and winter, attracted by a free-flowing stretch of the Chilkat River. Spawned-out salmon provide plenty of "fast food" for the birds. Their numbers peak in mid-November, but a few hundred inhabit the refuge year-round.

Glacier Bay National Park and Preserve

P.O. Box 140; Gustavus, AK 99826; tel: 907-697-2230.

In 1879, John Muir called Glacier Bay an "icy wilderness unspeakably pure and sublime." The park is 60 miles northwest of Juneau in Alaska's Panhandle and is reached only by boat or plane. Explore the bay by kayak or tour boat for views of humpback whales, orcas, sea otters, and harbor seals, as well as spectacular tidewater glaciers. On shore, look for brown bears, black bears, and mountain goats.

Wrangell-St. Elias National Park and Preserve

P.O. Box 29; Glennallen, AK 99588; tel: 907-822-5234.

"North America's Mountain Kingdom" encompasses four major mountain ranges and six of the continent's 10 highest peaks, including 18,000-foot Mount St. Elias, the second highest in Alaska. Wrangell-St. Elias is the size of six Yellowstones. It's the largest unit in the National Park System and, together with neighboring Kluane Park Preserve in Canada and Glacier Bay National Park, it forms the world's largest international preserved wilderness. Two unpaved roads penetrate the park. Both lead to historic mining towns and backcountry trails. River running and sea kayaking are also possible. Otherwise, access is limited to bush planes.

Denali
National Park
Alaska

CHAPTER
21

A cold wind rattles through a valley in **Denali National Park**. Gray skies the color of a lynx drop low and snow falls on high peaks, letting everyone know that this is the rooftop of the world. Birch leaves bright as the amber sap that flowed last May begin to vanish from the branches. A yellow-toothed beaver swims from its lodge through a thin sheet of ice forming on the water. Shuffling restlessly along the riverbank, a wolverine searches the sun-bleached driftwood piles for old salmon. Arctic ground squirrels burrow deeper into their dens as the sound of hooves comes from the hill across the river. ◆ First three caribou, then a dozen, crest the hill and move steadily toward the river. The caribou are the gray of twilight, with white manes and hocks. As the herd presses forward, the bulls hold their heads high, balancing the dra-matic spread of backswept antlers. The cows, with their smaller antlers, never stray far from their calves. Without pausing at the river-bank, the caribou swim through the water, powered by strong legs, their heads and shoulders held high. Without stopping, they trot in a smooth, even gait over the next tundra hill, disappearing as quickly as they came, following a path known only to them, a journey that will take them to the grounds of the autumnal rut. ◆ Above and just beyond the valley of the caribou rises snow-covered Denali, "The High One," whose deep glaciers of ice have helped create the rivers and shape the landforms that support the caribou and other legendary wildlife of the region. Denali, or Mount McKinley as it is also known, rises 20,320 feet at the center of Denali National Park, one of the largest national parks and most impressive wildlife sanctuaries in

Remote and wild, this legendary national park is the territory of grizzlies, caribou, wolves, and Dall sheep.

Denali, the Athabascan Indian name for Mount McKinley, presides over a landscape that supports almost 40 species of mammals, many of which can readily be spotted by visitors.

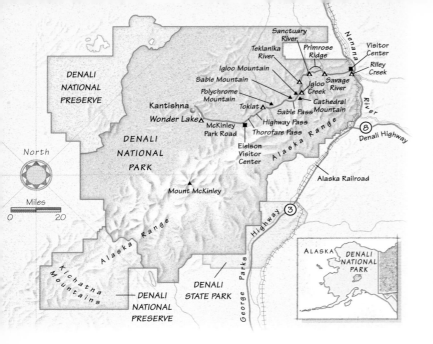

DENALI NATIONAL PRESERVE

Sanctuary River
Teklanika River
Primrose Ridge
Visitor Center
Riley Creek
Igloo Mountain
Sable Mountain
Igloo Creek
Savage River
Polychrome Mountain
Cathedral Mountain
Kantishna
Toklat
Sable Pass
Wonder Lake
McKinley Park Road
Highway Pass
Thorofare Pass
Denali Highway
8
DENALI NATIONAL PARK
Eielson Visitor Center
Alaska Range
North
Alaska Railroad
Mount McKinley
Miles
0   20
Alaska Range
Highway
3
ALASKA
DENALI NATIONAL PARK
Kichatna Mountains
DENALI STATE PARK
George Parks Highway
DENALI NATIONAL PRESERVE
Nenana River

the world. The park that surrounds the highest mountain in North America was formed in 1917, primarily through the efforts of hunter-naturalist Charles Sheldon, and was significantly enlarged in 1980. It and adjacent Denali National Preserve, established in 1980, stretch more than 100 miles along the Alaska Range and protect an area about the size of Massachusetts.

Summer Splendors

North of Denali and its satellite peaks, a 90-mile gravel road runs east-west through the fabled northern range of the park. Since access to this mecca for wildlife watchers is carefully regulated, most visitors concentrate their activities along the road. Beginning at **Riley Creek** near the park headquarters in the east, the rough track soon rises clear of the last bit of timber and rambles across the Alaskan tundra for over 50 miles. Because there are no trees, you have an essentially unobstructed view of grizzly bear, subarctic wolf, barren-ground caribou, Dall sheep, moose, and other wildlife. Along the way, the park road crosses five river valleys – the Savage, Sanctuary, Teklanika, East Fork of the Toklat, and Toklat – and climbs four mountain passes – the Sable, Polychrome, Highway, and Thorofare. At road's end is **Wonder Lake** and the former mining camp of **Kantishna** (now a wilderness resort area). The use of private automobiles on the park road is restricted after mile 15 at the **Savage River Check Station**, so visitors take tour or shuttle buses (riders are free to disembark at any point), hike into the park on foot, or ride mountain bicycles along the road.

Most wildlife enthusiasts come to Denali from late May through early September. Each

Barren-ground caribou (opposite), adept at swimming and traveling across rugged terrain and bogs, continually move throughout the park in search of food.

Wolves (right), like grizzly bears, are opportunistic hunters that prey on young moose, sheep, and caribou.

Mallards (below), green-winged teals, and widgeons, as well as shorebirds, are summertime residents of lakes and streams. More than 150 bird species have been recorded in the park.

month in the busy summer season presents its own wildlife-watching opportunities. June is best for viewing big predators such as wolves and grizzlies as they feed upon the young of the moose, Dall sheep, and caribou. It is not unusual for visitors to see animals killed within sight of the road and the carcasses defended by wolves or grizzlies. June is the peak breeding season for grizzlies, and large male bears sometimes venture into the road corridor, ordinarily the exclusive domain, or "nursery habitat," of females and their offspring. Another benefit is the midnight sun of the solstice period. Alaska at the latitude of Denali experiences nearly 24 hours of daylight in June, which means that energetic visitors can squeeze nearly two days of wildlife viewing into every day.

Caribou and moose are commonly seen near the park road from mid-June to late July. Dall sheep often cross the road on the way to their summer range. August is a favored time for Alaskans to visit the park, as the tundra colors begin their fabulous show after the first heavy frosts, usually around August 10. Caribou and moose antlers are fully formed by the middle of August, though they are still in velvet, and the blueberries come into season. The ripe berries attract grizzly bears to the road corridor, where they feed voraciously. Normally the northern lights can be seen after August 20, when

nighttime darkness begins to return to the sky. Heavy, road-closing snows can fall as early as Labor Day.

Land of Wolves and Dall Sheep

A short walk along the **Savage River** (mile 15 on the park road) offers an ideal introduction to the landscape and wildlife of Denali. Hikers soon notice that with each step they sink to their ankles in a spongy carpet of reindeer moss. This dense, moist layer of vegetation covers the underlying permafrost – literally, a permanent layer of ice beneath the living part of the ground. Once the art of navigating across moving ground is mastered, the journey begins. If it is fall, the berry crop on the tundra hills, from cranberries to blueberries, all in grocery-store profusion, will overwhelm. Wolves often leave their tracks in the soft sand and mud along the river. The far-ranging packs course through the broad treeless valley at dawn and dusk, coming from or going to their dens in the high country to the south.

At **Primrose Ridge**, directly west of the Savage River, visitors can journey deeper into the Denali backcountry. The climb to the summit is steep and not as easy as it looks from the road. The ridge soars half a mile into the sky, and most of the ascent is on a pitched slope and over wet tundra. There are no developed trails, and, as is the case throughout Denali, hikers navigate by line of sight. About halfway up, just past the rock outcroppings of Precambrian schist, the wet tundra no longer finds purchase on the slope. The rest of the ascent covers dry solid tundra, which is much easier on the ankles and knees. Signs of wildlife in this area include everything from the daybeds of

River of Bears

From late June through July, hundreds of thousands of migrating sockeye salmon converge on Brooks River in Katmai National Park. Midway up the river, the thronging salmon encounter Brooks Falls, a six-foot waterfall. As the salmon mass beneath the falls, they find themselves the prey of coastal brown bears.

Each six-pound salmon contains about 4,600 calories, most of it fat, which helps the bears bulk up for the coming winter. Up to 30 brown bears feed at one time at Brooks Falls, one of the great assemblages of the species on earth. The brown bears of Brooks Falls are among the largest on earth – larger than the interior grizzlies of Denali National Park – with adults regularly attaining weights of 1,500 pounds and more.

Observing and photographing the bears are primarily done on an elevated viewing platform on the south bank of Brooks River. Visitors can watch bears fishing (bears catch about one salmon for every half hour at the falls), fighting over choice fishing spots (bears live in a complex society based on dominance and size), nursing cubs, playing (play is considered by mammalogists as a sign of intelligence in animals), and courting and breeding (through early July). Bears can also be seen on the lower river from a second platform.

More than 12,000 people from around the world annually visit Brooks Falls and nearby Brooks Camp, including visitors who fly in for the day. So successful is the program at Brooks River that no person has ever been seriously injured by a bear, despite the regular close proximity. Brooks Camp also provides bear watching in September and October, after the peak berry season of August, as bears feed on spawned salmon that collect in slow water and along the banks. There are far fewer visitors than in summer, and the autumn colors can be spectacular.

Brown bears (left and center) station themselves at the base of Brooks Falls and jockey for prime fishing sites in skirmishes that wildlife watchers safely view from an observation platform.

McNeil River (below) is another well-known spot for watching bears. The river and its falls, west of the Brooks on the Alaska Peninsula, are protected in a sanctuary that visitors can enter by obtaining a permit.

Campers atop Ruth Glacier (opposite) enjoy spectacular mountain vistas. Visitors can hop in a bush plane for a stunning flight around Mount McKinley and then land on the glacier for hiking or camping.

grizzly bears (dug into the tundra like bathtubs to conceal the bears) to the close-cropped pastures of the local hoary marmots (which live in colonies on the rocky slopes of talus slides).

The top of Primrose Ridge is the exclusive home of the Dall sheep. These are, for many wildlife enthusiasts, the most beautiful sheep in the world, with pure white coats, compact muscular forms, and wide-curling golden horns. Among the rocky pinnacles and permanent snowfields of the alpine ridge, the hardy bands of Dall sheep thrive, relatively secure from wolves. Throughout the summer, the summit region is inhabited by ewes, lambs, and immature rams. Approach them quietly, maintaining a respectful distance, and these sheep may permit you to stay in their company for hours. It is always a delight to watch the young ones play, mimicking the behaviors they observe in adults, such as head butting or lip curling. In the autumn, the mature rams migrate back over from the glacial peaks to the south, where they spent the summer in "bachelor bands." By October, the breeding battles begin – bloodless contests in which the rams square off and butt horns in tremendous cracks that echo among the crags.

Home of the Grizzly

About 20 miles farther into the park, the old gravel road begins its long climb from the bottom of **Igloo Canyon**. To the north of the road rises **Sable Mountain**, a dark massive volcanic remnant. South of the road towers beautiful **Cathedral Mountain**, so named by pioneering explorer Charles Sheldon for its resemblance to a Gothic church. For the next five miles, off-road hiking is restricted, for this is the finest place in the park to observe grizzly bears. It is not uncommon for visitors to spot two or three family groups, numbering over 10 bears, along this stretch of road, particularly during the early summer when adult bears are attracted to such vegetation as Eskimo potato and bear-

flower, and during the early autumn, when bears find abundant blueberries and soapberries on the tundra.

Those who spend any time on **Sable Pass** soon learn that much of the mythology about the species is untrue, as grizzly bears are observed to be largely vegetarian (grazing on new grass like cattle), highly social, and quite playful. It is quite entertaining, for example, to watch a mother grizzly sliding down a snowbank on her head, her feet up in the air as her cubs look on, or two full-grown bear siblings, recognizing one another, wrestle and box over the tundra as if in the center ring of a circus tent. Grizzly fur can usually be found on the Sable Pass road sign at the top of the pass. Bears like to bite and claw the wooden sign, as well as rub their winter coats off against it. The

Dall sheep (above) dwell at the park's higher elevations. To find them, scan the mountains with the naked eye or with binoculars and look for white spots.

Antlers (left) found on the northern tundra are all that remain of a caribou that failed to survive a Denali winter.

silver-tipped fur feels long and luxurious, as smooth as corn silk (and often the same color), and always exudes an unmistakable ursine scent.

The tundra of Sable Pass and its environs (Sable Mountain and Cathedral Mountain) is also a fine place for birding. In many ways, the smooth rounded hills devoid of trees and rocks and the long brushy drainages of the high open country evoke the prairie. This is the domain of the golden eagle, gliding in the low arctic light as if in a watercolor by John James Audubon, and the long-tailed jaeger, with its black cap and white throat and spliced, knifelike tail. Here, too, is the ground-nesting lapland longspur, with its high-pitched song, *see, serilee-aw, serilee-ee, serilee-aw,* and the always-vigilant flocks of ravens with their cold metallic cries. For birders, Sable Pass is also a place of incongruities. A small group of mew gulls, which probably winter in Cook Inlet near Anchorage, regularly appear in spring and spend summer on the surrounding high tundra, scavenging wolf and grizzly kills.

Seeking Solitude

Ask sourdoughs (veteran Alaskans) about their favorite place in Denali National Park and they will invariably respond "Wonder Lake!" Situated at the end of the park road, this unusually large and deep lake (three miles long and shaped like a silver salmon) is a natural magnet for people and for wildlife. **Wonder Lake** is also one of the most beautiful lakes in the world, with the broad tundra domes rising all around and the great white mass of Denali perfectly reflected in its cold, cobalt-blue waters. Walking along the shore, one soon enters another world – of quiet spruce-covered islands and intimate coves where moose feed undisturbed among water lilies. This is the kingdom of the Arctic tern and the common loon, the beaver and the marten. If Denali has a peaceful center, a spiritual locus, it is in this restful body of water.

Beyond Wonder Lake is the roadless, trackless Alaskan wilderness. These thousands of square miles of unknown forests

and unnamed lakes are familiar only to the grouse and the goshawk, the lynx and the fox. This vast frontier, much of it protected as **Denali National Preserve**, can be reached by bush plane. The pilots, pioneers of the outback, take wilderness lovers into remote areas where wildlife can be viewed in complete solitude. From the backseat of a bush plane, you can take in the country as a unified whole – the thousands of lakes, the wide-coursing rivers, the extraordinary presence of Denali at the center of it all. Here is the lost north country of Jack London and Robert Service, of grizzled prospectors and frost-bitten explorers. Here are the lonely steppes and rivers that the first Asian immigrants saw, tens of thousands of years ago, as they began to explore the New World – a land that is as wild and pristine now as it was then.

Moose, well adapted for life in the far north, have a thick coat that warms them in winter and protects them from summer's biting insects.

TRAVEL TIPS

DETAILS

When to Go

Peak visitation occurs from late May to mid-September. Summer is cool and wet, with temperatures ranging from 40° to 70°F. Winter temperatures range from 40° to well below zero. Spring is brief and usually cool; snow covers the ground until mid-May. Fall is also brief, with autumn colors peaking in early September. High winds and storms occur year-round.

How to Get There

Car rentals are available at Anchorage International and Fairbanks International Airports. Rail, bus, and charter services are available from Fairbanks and Anchorage. The park is 237 miles north of Anchorage and 120 miles south of Fairbanks on Alaska Highway 3.

Getting Around

Travel on the park road is limited to shuttle buses, tour buses, bicycles, and hiking. Shuttle buses will drop off or pick up anywhere along the route. Bus tickets are limited. Reservations should be made well in advance of your visit. For information or reservations, call 800-622-7275 (for shuttle) or 800-276-7234 (for tour bus).

Backcountry Travel

A backcountry use permit, required for overnight camping, must be obtained in person at the visitor center in summer and at park headquarters in winter. A quota system is enforced.

Handicapped Access

The visitor center, Savage Cabin Trail, restrooms, and three campgrounds are accessible.

INFORMATION

Denali National Park and Preserve

P.O. Box 9; Denali Park, AK; 99755-0009; tel: 907-683-2294.

Alaska Tourism

P.O. Box 110801; Juneau, AK 99811-0801; tel: 907-465-2010.

Fairbanks Visitor Information Center

550 First Avenue, Suite 306; Fairbanks, AK 99701; tel: 800-327-5774 or 907-456-5774.

CAMPING

The park has seven campgrounds with a total of 293 sites, many of which may be reserved in advance. To reserve a campsite, call 800-622-7275 or write to Denali Park Resorts VTS; 241 West Ship Creek Avenue; Anchorage, AK 99501. Mail requests must be received 30 days prior to visit. Most campgrounds are open from May to September. One campground, Riley Creek, is open year-round.

LODGING

PRICE GUIDE – double occupancy

$ = up to $49 $$ = $50-$99
$$$ = $100-$149 $$$$ = $150+

Camp Denali

P.O. Box 67; Denali National Park, AK 99755; tel: 907-683-2290.

Located in the remote Kantishna area, at the end of the park's 93-mile road, Camp Denali was founded in 1952 by two women bush pilots. The camp's 17 log cabins command striking views of Mount McKinley; they are privately located and simply furnished, each with a wood-burning stove and outhouse. A modern bathroom and shower facility is situated nearby. Prices include meals, served in the historic log lodge. The Natural History Resource Center offers wide-ranging exhibits, lectures, and week-long programs. Multi-day stays only. Amenities include a library, photography darkroom, canoes, and bicycles. Open early June to early September. $$$$

Denali Backcountry Lodge

P.O. Box 189; Denali Park, AK 99755; tel: 907-683-1341 (June 1 to September 15) or P.O. Box 810; Girdwood, AK 99587; tel: 907-783-1342 (September 16 to May 31).

This lodge is in Kantishna and offers 30 comfortably appointed cabins with private baths. Prices include meals and a shuttle bus. Guided activities and naturalist programs are also available. Open June to September. $$$$

Denali Crow's Nest

P.O. Box 70; Denali National Park, AK 99755; tel: 907-683-2723.

On Sugar Loaf Mountain one mile north of the park entrance, these 39 log cabins overlook Horseshoe Lake, the Nenana River, and the Alaska Range. Each unit can accommodate up to four people in basic comfort, and includes two double beds, a private bath, and deck. Extras include free shuttle service, hot tub, restaurant, and booking service for guided tours and activities. Open late May to early September. $$$–$$$$

Denali Windsong Lodge

P.O. Box 221011; Anchorage, AK 99522; tel: 800-208-0200 or 907-245-0200.

This lodge is set in a quiet spot about a mile from the park entrance. Built in 1991 and enlarged in 1996, Denali Windsong offers 72 rooms, all with great mountain views, private entrances, two double beds, full baths, and log furniture. $$–$$$$

North Face Lodge

P.O. Box 67; Denali National Park, AK 99755; tel: 907-683-2290.

Built in 1973, this lodge is located on tundra meadow in the Moose Creek Valley, a mile from its sister lodge, Camp Denali. Fifteen guest rooms are modestly appointed, each with private bath. Mount McKinley is visible from the lodge's large patio. Guests participate in a variety of active learning programs. A library, photography darkroom, canoes, and bicycles are also included. Open June to early September. $$$$

TOURS & OUTFITTERS

Alaska Wildland Adventures

P.O. Box 389; Girdwood, AK 99587; tel: 800-334-8730.

Natural-history tours range from five to 12 days and feature rafting, kayaking, photography, and guided hikes. Lodging is provided.

Denali Air

P.O. Box 82; Denali National Park, AK 99755; tel: 907-683-2261.

Flightseeing tours depart from a private air strip in the Alaska Range. One-hour excursions fly within a mile of Mount McKinley. Charter flights are also available.

Denali Backcountry Guides

P.O. Box 540; Healy, AK 99743; tel: 907-683-2419

Day and overnight trips led by skilled naturalists and mountaineers, with an emphasis on nature photography.

Denali Raft Adventures

Drawer 190; Denali National Park, AK 99755; tel: 888-683-2234 or 907-683-2234.

Overnight, all-day, two- and four-hour excursions on the Nehana River. Rafters often see moose, sheep, caribou, and bears. Several packages allow various levels of participation and whitewater.

Excursions

Arctic National Wildlife Refuge

101 12th Avenue; P.O. Box 20; Fairbanks, AK 99701; tel: 907-456-0250.

The "Serengeti of the North" lies above the Arctic Circle around the snowcapped Brooks Range. A surprising diversity of wildlife roams this harsh but fragile land. Musk oxen graze on tundra plants, gyrfalcons patrol the skies, grizzlies dig for ground squirrels, and wolves hunt the mountains. The most spectacular event, however, is the migration of countless thousands of caribou to and from their calving grounds near the coast. Visitors must be completely self-sufficient and experienced in wilderness travel.

Lake Clark National Park and Preserve

4230 University Drive, Suite 311; Anchorage, AK 99508; tel: 907-271-3751.

Lake Clark is a vignette of wild Alaska just an hour by plane from Anchorage You'll find a little bit of everything here – sawtooth mountains, forested coast, pristine tundra, crystal lakes, and two active volcanoes. Wildlife is equally diverse. Moose feed in ponds and streams, grayling fill the rivers, Dall sheep climb into the highlands, and bears are just about everywhere. Hiking is rugged. Kayaking is easier and allows more freedom of movement.

Yukon Delta National Wildlife Refuge

P.O. Box 346; Bethel, AK 99559; tel: 907-543-3151.

Millions of waterfowl, shorebirds, and passerines migrate to this huge preserve on the Bering Sea. Gyrfalcons, golden eagles, rough-legged hawks, and peregrine falcons nest here, as do 145 other species of birds. The refuge's varied habitat – tundra, marsh, grasslands, bogs – also sustains black and brown bears, caribou, moose, wolves, red and Arctic foxes, and wolverines. Beluga whales and walrus sometimes make their way up the river delta.

SECTION THREE

Resource Directory

FURTHER READING

Ecology & Environment

The Conservation of Whales and Dolphins: Science and Practice, Mark P. Simmonds and Judith D. Hutchinson, eds. (J. Wiley, 1996).

The Control of Nature, by John McPhee (Farrar, Straus, Giroux, 1989).

Ecologist: 20 Answered Questions for Busy People Facing Environmental Issues, by John Janovy, Jr. (St. Martin's Press, 1997).

The Ecology of Fishes on Coral Reefs, by Peter F. Sale, ed. (Academic Press, 1994).

The Ecology of North America, by Victor E. Shelford (University of Illinois Press, 1963).

The Ecology of Whales and Dolphins, by D.E. Gaskin (Heinemann, 1982).

A Fierce Green Fire: The American Environmental Movement, by Philip Shabecoff (Hill and Wang, 1993).

Keeping All the Pieces: Perspectives on Natural History and the Environment, by Whit Gibbons (Smithsonian, 1993).

The Klamath Knot, David Rains Wallace (Random House, 1984).

Life in the Balance, by David Rains Wallace (Harcourt Brace Jovanovich, 1987).

Our Endangered Parks, by National Parks and Conservation Association (Foghorn Press, 1994).

The Quiet Crisis, by Stuart Udall (Peregrine Smith Books, 1988).

Saving America's Wildlife, by Thomas R. Dunlap (Princeton University, 1991).

Silent Spring, by Rachel Carson (Houghton Mifflin, 1962).

The Web of Life, by John H. Storer (Devin-Adair, 1953).

Wildlife in America, by Peter Matthiessen (Viking Press, 1959).

Nature

American Nature Writing 1998, by John A. Murray, ed. (Sierra Club Books).

Among Whales, by Roger Payne (Scribner, 1995).

Arctic Dreams: Imagination and Desire in a Northern Landscape, by Barry Holstun Lopez (Bantam Books, 1996).

Birch Browsings: A John Burroughs Reader, by Bill McKibben, ed. (Penguin, 1992).

The Buffalo Book: The Full Saga of the American Animal, by David Dary (Swallow, 1989).

Cactus Country: An Illustrated Guide, by John A. Murray (Roberts Rinehart Press, 1997).

A City Under the Sea: Life in a Coral Reef, by Norbert Wu (Atheneum, 1996).

Crossing Open Ground, by Barry Holstun Lopez and Marty Asher (Vintage Books, 1989).

The Desert Bighorn: Its Life History, Ecology, and Management, by Gale Monson and Lowell Sumner, eds. (University of Arizona Press, 1981).

Desert Notes: Reflections in the Eye of a Raven, by Barry Holstun Lopez (Avon, 1981).

Desert Solitaire: A Season in the Wilderness, by Edward Abbey (Ballantine, 1968).

The Desert Year, by Joseph Wood Krutch (Viking, 1952).

Eyes Open in the Dark: Eight Essays, by Conger Beasley, Jr. (BkMk Press, 1996).

The Great Bear Almanac, by Gary Brown (Lyons & Burford, Publishers, 1993).

The Immense Journey, by Loren Eiseley (Random House, 1957).

The Maine Woods, by Henry David Thoreau (Penguin Books, 1988).

Mammals of the American North, by Adrian Forsyth (Camden House, 1985).

The Natural History Essays, by Henry David Thoreau (Gibbs Smith, 1980).

The Norton Book of Nature Writing, by Robert Finch and John Elder, eds. (Norton, 1990).

Pioneer Naturalists: The Discovery and Naming of North American Plants and Animals, by Howard Ensign Evans (Holt, 1993).

A Sand County Almanac, by Aldo Leopold (Oxford University Press, 1948).

A Sharp Lookout: Selected Nature Essays of John Burroughs, by Frank Bergon, ed. (Smithsonian, 1987).

South of Yosemite: Selected Writings of John Muir, by John Muir (Wilderness Press, 1988).

The Star Thrower, by Loren Eiseley (Harcourt Brace, 1979).

This Incomperable Lande: A Book of American Nature Writing, by Thomas J. Lyon, ed. (Penguin, 1991).

Thoreau on Birds: Notes on New England Birds from the Journals of Henry David Thoreau (Beacon 1993).

Voices in the Desert: Writings and Photographs, by Larry Cheek (Harcourt Brace Children's Books, 1995).

Water and Sky: Reflections of a Northern Year, by Alan Kesselheim (Fulcrum, 1989).

Where the Bluebird Sings to the Lemonade Spring: Living and Writing in the West, by Wallace Stegner (Random House, 1992).

Wild Cats: Lynx, Bobcats, Mountain Lions, by Candace Sherk Savage (Sierra Club, 1993).

The Wind Masters: The Lives of North American Birds of Prey, by Peter Dunne (Houghton Mifflin Company, 1995).

Regional Titles

Adventure Guide to the Florida Keys & Everglades National Park, by Joyce Huber (Hunter Publishers, 1997).

Alaska's Bears: Grizzlies, Black Bears, and Polar Bears, by Bill Sherwonit (Alaska Northwest Books, 1998).

Alaska's Mammals: A Guide to Selected Species, by Dave Smith (Alaska Northwest Books, 1995).

America's Secret Recreational Areas: Our Guide to the Forgotten Wild Lands of the Bureau of Land Management, by Michael Hodgson (Foghorn Press, 1995).

Amphibians & Reptiles of Yellowstone and Grand Teton National Parks, by Edward D. Koch (University of Utah Press, 1995).

Arizona Day Hikes: A Guide to the Best Trails from Tucson to the Grand Canyon, by Dave Ganci (Sierra Club Books, 1995).

Badlands, Theodore Roosevelt, and Wind Cave National Parks, by Michael Milstein (Northword Press, 1996).

The Bears of Yellowstone, by Paul D. Schullery (Harbinger House, 1992).

Birds of Yellowstone: A Practical Habitat Guide to the Birds of Yellowstone National Park – And Where to Find Them, by Terry McEneaney (Roberts Rinehart Publishers, 1988).

Bird Studies at Old Cape May, 2 vols., by Witmer Stone (Dover Publications, 1965).

The Complete Guidebook to Yosemite National Park, by Steven P. Medley (Elizabeth O'Neill, 1998).

The Curious Country: Badlands National Park, by Mary Durant and Michael Harwood (Badlands Natural History Association, 1988).

Day Hikes in Yosemite National Park: 25 Favorite Hikes, by Robert Stone (Ics Books, 1997).

Discovering Yellowstone Wolves, by James C. Halfpenny (Falcon Press Publishing, 1996).

Everglades, by Connie Toops (Voyageur Press, 1989).

Exploring Beyond Yellowstone: Hiking, Camping, and Vacationing in the National Forests Surrounding Yellowstone and Grand Teton, by Ron Adkison (Wilderness Press, 1996).

Exploring the Yellowstone Backcountry: A Guide to the Hiking Trails of Yellowstone With Additional Sections on Canoeing, Bicycling, and Cross-country, by Orville E. Bach, Jr. (Sierra Club Books, 1998).

Exploring Wild South Florida: A Guide to Finding Natural Areas and Wildlife of the Southern Peninsula and the Florida Keys, by Susan D. Jewell (Pineapple Press, 1997).

50 Best Short Hikes in Yosemite and Sequoia/Kings Canyon, John Krist (Wilderness Press, 1993).

50 Hikes in New Jersey: Walks, Hikes, and Backpacking Trips from the Kittatinnies to Cape May, by Bruce C. Scofield, Stella J. Green, and H. Neil Zimmerman (Backcountry Publications, 1997).

Florida Wildlife Viewing Guide, by Susan Cerulean and Ann Morrow (Falcon Publishing Company, 1998).

The Forgotten Peninsula: A Naturalist in Baja California, by Joseph Wood Krutch (University of Arizona Press, 1986).

Great Smoky Mountains National Park, by Rose Houk (Houghton Mifflin Company, 1993).

Great Smoky Mountains National Park: Wildlife Watcher's Guide, by Mike Carlton (Northwood Press, 1996).

The Guide to National Parks of the Southwest, by Nicky Leach (Southwest Parks & Monuments Association, 1992).

Guide to National Wildlife Refuges, by Laura and William Riley (Collier Books, 1992).

In Denali, by Kim Heacox (Companion Press, 1992).

Introduction to Grand Canyon Ecology, Rose Houk (Grand Canyon Association, 1997).

Kenai Pathways: A Guide to the Outstanding Wildland Trails of Alaska's Kenai Peninsula (Alaska Natural History Association, 1995).

Mountain Islands and Desert Seas: A Natural History of the U.S.-Mexico Borderlands, by Frederick R. Gehlbach (Texas A&M University, 1993).

Natural History of Baja California, by Miguel Del Barco (Dawsons Book Shop, 1980).

The Natural History of Big Sur, by Paul Henson and Donald J. Usner (University of California, 1993).

The Nature of Southeast Alaska: A Guide to Plants, Animals, and Habitats, by Rita M. O'Clair (Graphic Arts Center Publishing Company, 1997).

Nature Travel, by Dwight Holing *et al.* (The Nature Company, 1995).

Nebraska Wildlife Viewing Guide, by Joseph Knue (Falcon Publishing Company, 1997).

Olympic National Park: Where the Mountain Meets the Sea (Woodlands Press, 1984).

Portrait of Alaska's Wildlife, by Tom Walker (Graphic Arts Center Publishing, 1997).

Reflections from the North Country, by Sigurd F. Olson (Alfred A. Knopf, 1976). Observations on Minnesota's environment.

Saguaro: A View of Saguaro National Monument & the Tucson Basin, by Gary Paul Nabhan (SW Parks & Monuments Association, 1986).

Sea of Cortez Marine Animal: A Guide to Common Fishes and Invertebrates, Baja California to Panama, by Daniel Gotshall (Sea Challengers, 1998).

Southcentral Alaska: Including Anchorage, Kenai Peninsula, Susitna Valley and Prince William Sound, by Scott McMurren (Epicenter Press, 1995).

U.S. National Parks East, by John Gattuso, ed. (Apa Publications, 1995).

U.S. National Parks West, by John Gattuso, ed. (Apa Publications, 1995).

Washington Wildlife Viewing Guide, by Joe La Tourrette (Falcon Publishing Company, 1992).

Wild Encounters: The Best Animal-Watching Adventures in the U.S., by Diane Bair and Pamela Wright (Willow Creek Press, 1998).

Wind Cave: An Ancient World Beneath the Hills, by Arthur N. Palmer (Wind Cave/Jewel Cave Natural History Association, 1988).

Yellowstone and Grand Teton National Parks: A Traveler's Guide, by Steven Fuller and Jeremy Schmidt (Free Wheeling Travel Guides, 1991).

Natural History & Field Guides

All About Saguaros, by Carle Hodge (Arizona Highways, 1997).

Birding, by Terence Lindsey, ed. (Time-Life Books, 1994).

Birds: An Explore Your World Handbook (Discovery Communications, Inc., 1999)

The Complete Tracker: Tracks, Signs, and Habits of North American Wildlife, by Len McDougall (The Lyons Press, 1997).

A Field Guide to Animal Tracks, by Olaus J. Murie (Houghton Mifflin Company, 1972).

Field Guide to Coral Reefs: Caribbean and Florida, by Eugene H. Kaplan (Houghton Mifflin Company, 1988).

A Field Guide to Reptiles & Amphibians: Eastern and Central North America, by Roger Conant and Joseph T. Collins (Chapters Publications Ltd., 1998).

A Field Guide to the Birds: A Completely New Guide to All the Birds of Eastern and Central North America, by Roger Tory Peterson (Houghton Mifflin Company, 1998).

A Field Guide to the Ecology of Western Forests, by John C. Kricher (Houghton Mifflin Company, 1993).

A Field Guide to the Mammals: North America North of Mexico, by William H. Burt and Richard P. Grossenheider (Chapters Publications Ltd., 1998).

A Field Guide to Western Birds: A Completely New Guide to Field Marks of All Species Found in North America West of the 100th Meridian and North of Mexico, by Roger Tory Peterson (Houghton Mifflin Company, 1998).

A Guide to Animal Tracking and Behavior, by Donald W. Stokes (Little Brown & Company, 1987).

Grizzly Bears: An Illustrated Field Guide, by John A. Murray (Roberts Rinehart Press, 1995).

Hawks in Flight: The Flight Identification of North American Migrant Raptors, by Peter Dunne (Houghton Mifflin Company, 1989).

Hawk Watch: A Guide for Beginners, Peter Dunne (New Jersey Audubon Society, 1990).

Living with Wildlife: How to Enjoy, Cope with, and Protect North America's Wild Creatures Around Your Home and Theirs, by the California Center for Wildlife (Sierra Club Books, 1994).

Mammals of the Northern Rockies, by Tom J. Ulrich (Mountain Press Publishing, 1990).

National Audubon Society First Field Guide: Mammals, by John Grassy and Chuck Keene (Chanticleer Press, 1998).

National Audubon Society Field Guide to North American Birds: Eastern Region, by John Bull and John Farrand, Jr. (Knopf, 1994).

National Audubon Society Field Guide to North American Birds: Western Region, by Miklos D. F. Udvardy and John Farrand, Jr. (Knopf, 1994).

National Audubon Society Field Guide to North American Fishes, Whales, and Dolphins, by Herbert T. Boschung and David K. Caldwell (Knopf, 1983).

National Audubon Society Field Guide to North American Mammals, by John O. Whitaker, Jr. (Knopf, 1996).

National Audubon Society Field Guide to North American Reptile and Amphibians, by Roger Tory Peterson, F. Wayne King, and John L. Behler (Knopf, 1979).

National Audubon Society Field Guide to Tropical Marine Fishes: Of the Caribbean, the Gulf of Mexico, Florida, the Bahamas, and Bermuda, by C. Lavett Smith (Knopf, 1997).

The Sierra Club Handbook of Seals and Sirenians, by B. Stewart (Sierra Club Books, 1992).

The Sierra Club Handbook of Whales and Dolphins, by Stephen Leatherwood and Randall R. Reeves (Sierra Club Books, 1993).

Stokes Field Guide to Birds: Eastern Region, by Donald W. Stokes and Lillian Q. Stokes (Little, Brown and Company, 1996).

Suburban Nature Guide: How to Discover and Identify the Wildlife in Your Backyard, by David Mohradt and Richard E. Schinkel (Stackpole Books, 1991).

Tracking and the Art of Seeing, by Paul Rezendes (Camden House Publishing, 1992).

The Walker's Companion, by David Rains Wallace, ed. (Time-Life Books, 1995).

Watching Nature: A Beginner's Field Guide, by Monica Russo (Sterling Publications, 1998).

Watching Wildlife: Tips, Gear and Great Places for Enjoying America's Wild Creatures, by Mark Damian Duda (Falcon Publishing Company, 1995).

Whales, Dolphins & Porpoises, by Mark Carwardine and Erich Hoyt, eds. (Time-Life Books, 1998).

The Whale Watcher's Guide, by Patricia Corrigan (Northword Press, 1994).

Preparation & Equipment

The Cordes/LaFontaine Pocket Guide to Outdoor Photography, by Mary Mather (Graycliff Publishing, 1994).

Easy Access to National Parks: The Sierra Club Guide for People with Disabilities, by Wendy Roth (Sierra Arts Foundation, 1992).

The Essential Guide to Hiking in the United States, by Charles Cook (Michael Kescend Publishing, 1992).

The Explorer Naturalist, by Vinson Brown (Stackpole Books, 1976).

Loving Nature The Right Way: A Family Guide to Viewing and Photographing Scenic Areas and Wildlife, Bruce Hopkins (Partnership Press, 1997).

Making Camp: A Complete Guide for Hikers, Mountain Bikers, Paddlers & Skiers, by Steve Howe, Alan Kesselheim, Dennis Coello, and John Harlin (The Mountaineers, 1997).

Night Life: Nature from Dusk to Dawn, by Diana Kappel-Smith (Little, Brown, 1990). How to study nature at night.

Photography of Natural Things, by Freeman Patterson (Sierra Club Books, 1990).

The Sierra Club Family Outdoors Guide, by Marilyn Doan (Sierra Club Books, 1995).

The Sierra Club Guide to Close-Up Photography in Nature, by Tim Fitzharris (Sierra Club Books, 1998).

The Sierra Club Guide to 35Mm Landscape Photography, by Tim Fitzharris (Sierra Club Books, 1996).

Walking Softly in the Wilderness: The Sierra Club Guide to Backpacking, by John Hart (Sierra Club Books, 1994).

Wilderness Basics: The Complete Handbook for Hikers and Backpackers, by Jerry Schad (The Mountaineers, 1993).

Magazines

Audubon
National Audubon Society, 950 Third Avenue; New York, NY 10022.

Backpacker
Rodale Press, 33 East Minor Street; Emmaus, PA 18098.

Explore
301 14th Street NW, Suite 420; Calgary, Alberta, T2N 2A1.

National Parks
National Parks and Conservation Association, 1701 18th Street NW; Washington, DC 20009.

National Wildlife
National Wildlife Federation, 1400 16th Street NW; Washington, DC 20036.

Natural History
American Museum of Natural History, Central Park West at 79th Street; New York, NY 10024.

Nature Conservancy
The Nature Conservancy, 1815 North Lynn Street; Arlington, VA 22209.

Outdoor Photographer
12121 Wilshire Boulevard, Suite 1220; Los Angeles, CA 90025-1175.

Outdoor Traveler
WMS Publications, One Morton Drive, Suite 102; Charlottesville, VA 22903.

Outside
Outside Plaza; 400 Market Street; Santa Fe, NM 87501.

Sierra
The Sierra Club, 730 Polk Street; San Francisco, CA 94109.

Wilderness
The Wilderness Society, 900 17th Street NW; Washington, DC 20006-2300.

ORGANIZATIONS

Alaska Natural History Association
P.O. Box 230; Denali National Park, AK 99755; tel: 907-683-1258.

American Birding Association
P.O. Box 6599; Colorado Springs, CO 80225; tel: 800-835-2473.

American Cetacean Society
P.O. Box 1391; San Pedro, CA 90731; tel: 310-548-6279.

American Hiking Society
P.O. Box 20160; Washington, DC 20041-2160; tel: 703-319-0084.

American Wilderness Experience
P.O. Box 1468; Boulder, CO 80306; tel: 800-444-0099.

Appalachian Mountain Club
P.O. Box 298, Route 16; Gorham, NH 03581; tel: 603-466-2727.

Badlands Natural History Association
P.O. Box 6; Interior, SD 57750; tel: 605-433-5361.

Big Bend Natural History Association
P.O. Box 68; Big Bend National Park, TX 79834; tel: 915-477-2236.

Bryce Canyon Natural History Association
Bryce Canyon, UT 84717; tel: 801-834-5322.

Canyonlands Natural History Association
30 South 100 East; Moab, UT 84532; tel: 801-259-6003.

Carlsbad Caverns-Guadalupe Mountains Association
P.O. Box 1417; Carlsbad, NM 88221; tel: 505-785-2318.

Crater Lake Natural History Association
P.O. Box 157; Crater Lake, OR 97604; tel: 503-594-2211.

Death Valley Natural History Association
P.O. Box 188; Death Valley, CA 92328; tel: 619-786-2331.

Glacier Natural History Association
P.O. Box 428; West Glacier, MT 59936; tel: 406-888-5756.

Grand Canyon Natural History Association
P.O. Box 399; Grand Canyon, AZ 86023; tel: 602-638-2481.

Grand Teton Natural History Association
P.O. Box 170; Moose, WY 83012; tel: 307-739-3404.

Joshua Tree Natural History Association
74485 National Monument Drive; Twentynine Palms, CA 92277; tel: 619-367-1488.

National Audubon Society
950 Third Avenue; New York, NY 10022; tel: 212-832-3200.

National Campers and Hikers Association
4804 Transit Road, Building 2; Depew, NY 14043; tel: 716-668-6242.

National Parks and Conservation Association
1776 Massachusetts Avenue NW, Suite 200; Washington, DC 20036; tel: 202-797-6800.

National Recreation and Park Association
2775 South Quincy Street, Suite 300; Arlington, VA 22206-2204; tel: 703-671-6772.

National Wildlife Federation
1400 16th Street NW; Washington, DC 20036; tel: 202-223-6722.

The Nature Conservancy
1815 North Lynn Street; Arlington, VA 22209; tel: 703-841-5300.

North Cascades Institute
2105 Highway 20; Sedro Woolley, WA 98284; tel: 206-856-5700.

Northwest Interpretive Association
3002 Mount Angeles Road; Port Angeles, WA 98362; tel: 206-452-4501.

Olympic Park Institute
111 Barnes; Port Angeles, WA 98363; tel: 800-775-3720.

Point Reyes Natural History Association
Bear Valley Road; Point Reyes, CA 94956; tel: 415-663-1200.

Redwood Natural History Association
1111 2nd Street; Crescent City, CA 95531; tel: 707-464-9150.

Rocky Mountain Nature Association
Rocky Mountain National Park; Estes Park, CO 80517; tel: 303-586-1399.

Sequoia Natural History Association
HCR-89, Box 10; Three Rivers, CA 93271; tel: 209-656-3758.

Sierra Club
730 Polk Street; San Francisco, CA 94109; tel: 415-923-5630.

Southwest Parks and Monuments Association
221 North Court Avenue; Tucson, AZ 85701; tel: 602-622-1999.

Wilderness Society
900 17th Street NW; Washington, DC 20006; tel: 202-833-2300.

Yellowstone Association
P.O. Box 117; Yellowstone National Park, WY 82190; tel: 307-344-2293.

Yosemite Association
P.O. Box 230; El Portal, CA 95318; tel: 209-372-0420.

Zion Natural History Association
Springdale, UT 84767; tel: 800-635-3959.

GOVERNMENT AGENCIES

Bureau of Land Management
U.S. Department of the Interior, 1849 C Street NW; Washington, D.C. 20240; tel: 202-208-5717.

Fish and Wildlife Service
U.S. Department of the Interior, 1849 C Street NW; Washington, D.C. 20240; tel: 202-208-5717.

Forest Service
U.S. Department of Agriculture, 14th and Independence Avenue SW, S Agriculture Building; Washington, D.C. 20250; tel: 202-205-8333.

National Park Service
Office of Public Inquiries, P.O. Box 37127; Washington, D.C. 20013; tel: 202-208-4747.

TOURISM INFORMATION

Alaska Tourism
P.O. Box 110801; Juneau, AK 99811-0801; tel: 907-465-2010.

Arizona Office of Tourism
1100 West Washington Street; Phoenix, AZ 85007; tel: 800-842-8257 or 602-542-8687.

Arkansas Tourism
1 Capital Mall; Little Rock, AR 72201; tel: 501-682-1088.

Baja Information
7860 Mission Center Court No. 2; San Diego, CA 92108; tel: 800-225-2786 or 800-522-1516 (in California).

California State Division of Tourism
801 K Street, Suite 1600; Sacramento, CA 95814; tel: 800-462-2543 or 916-322-2881.

Florida Division of Tourism
126 West Van Buren Street; Tallahassee, FL 32399; tel: 904-487-1462.

Georgia Tourism
285 Peachtree Center Avenue, Suite 1000; Atlanta, GA 30303; tel: 800-847-4842 or 404-656-3590.

Maine Tourism
P.O. Box 2300; Hallowell, ME 04347; tel: 800-533-9595 or 207-623-0363.

Minnesota Tourism
500 Metro Square; 1217 Place; St. Paul, MN 55101; tel: 800-657-3700 or 612-296-5029.

Nebraska Travel and Tourism
P.O. Box 98907; Lincoln, NE 68509-8907; tel: 800-228-4307 or 402-471-3796.

New Jersey Travel and Tourism
20 West State Street; Box 826; Trenton, NJ 08628; tel: 800-537-7397.

North Carolina Division of Travel and Tourism
430 North Salisbury Street; Raleigh, NC 27611; tel: 919-733-4171.

South Dakota Department of Tourism
711 Wells Avenue; Pierre, SD 57501; tel: 605-773-3301.

Tennessee Tourism
320 Sixth Avenue North, 5th Floor; Nashville, TN 37219; tel: 800-836-6200 or 615-741-2159.

Washington State Tourism
P.O. Box 42500; Olympia, WA 98504-2500; tel: 800-544-1800 or 360-586-2088.

Wyoming Division of Tourism
I-25 at College Drive; Cheyenne, WY 82002; tel: 800-225-5996 or 307-777-7777.

PHOTO AND ILLUSTRATION CREDITS

Nancy Adams/Tom Stack & Associates 149

Mike Bacon/Tom Stack & Associates 49 (3rd from top), 56T

Craig Blacklock/Larry Ulrich Stock Photography, Inc. 12/13, 16, 28, 71B, 102, 105, 108T, 206T

Randy Brandon/AlaskaStock Images 47T

Dominique Braud/Tom Stack & Associates 24B, 37T, 111M, 125M

John Cancalosi/Tom Stack & Associates 49 (2nd from bottom), 115B, 144

Gerald and Buff Corsi 18B, 33B, 109T, 139

David M. Dennis/Tom Stack & Associates 32TL, 72B, 75T, 87T

Mark Dietz 120

Patrick J. Endres 25, 46/47, 195, 200, 207, 209M

Bill Everitt/Tom Stack & Associates 83B

Jeff Foott front cover, 6/7, 30B, 35, 86, 126, 134, 140, 143 (2nd from top), 146/147, 148T, 148B, 162B, 167B, 192, 204T

Jeff Foott/Tom Stack & Associates 36B, 48T, 130T, 170, 175T

Michael H. Francis 133B

Friends of the Everglades 89T

John Gerlach/Tom Stack & Associates 38B, 88, 104L

Francois Gohier 9B, 30T, 47B, 48 (3rd from top), 48B, 49B, 64, 133M, 143 (3rd from top), 152, 154T, 156B, 159M, 165L

Joe Mac Hudspeth, Jr. 37B, 49 (2nd from top), 81T, 89B, 112, 131

George H.H. Huey 5T, 31B, 39T, 151B, 159B, 166, 169T

Virginia Hurst/Tom Stack & Associates 111T

Gavriel Jecan/Art Wolfe, Inc. 26B, 189T

Johnny Johnson/AlaskaStock Images 199T

Lewis Kemper 10/11, 20/21, 138B, 143T, 175B

Thomas Kitchin/Tom Stack & Associates 34T, 41T, 91, 106B, 141B, 176T, 176B, 196B, 209T, back cover top

Lon Lauber/AlaskaStock Images 46T

Bill Lea 41B, 68, 72T, 77T, 77B, 78, 81B, 93T, 137BR, 143B

Tom and Pat Leeson 9T, 49 (3rd from bottom), 50/51, 84, 106T, 114, 115T, 123, 129T, 138T, 172, 179T, 185, 189M, 196T, 204M, 204B, 206B

Ron Levalley/Larry Ulrich Stock Photography, Inc. 157, 159T, 167T

Library of Congress 56B, 194T

Joanne Lotter/Tom Stack & Associates 187

Joe McDonald/Tom Stack & Associates 18T, 36T, 38T, 62, 70, 87B, 104R, 107

Colin McRae 8R, 27T

Randy Morse/Tom Stack & Associates 164

William Neill/Larry Ulrich Stock Photography, Inc. 4, 57, 61M, 74/75

Mark Newman/Tom Stack & Associates 205

Michele Nolan/Tom Stack & Associates 165R

Brian Parker/Tom Stack & Associates 83T, 98B, 101M, 101B

Doug Perrine/Mo Yung Productions 101T

Rob Planck/Tom Stack & Associates 39B, 116B

Cliff Reidinger/AlaskaStock Images 22

Carl R. Sams II 32TR, 34B, 67T, 71T, 73T, 73B, 93M, 116T, 119M

Gary Schultz/AlaskaStock Images 14/15

Jeff Schultz/AlaskaStock Images 26T

Wendy Shattil and Bob Rozinski/Tom Stack & Associates 48 (3rd from bottom)

John Shaw/Tom Stack & Associates 137BL, 183

Bill Sherwonit 141T

Bill Silliker, Jr. 52, 54, 55B, 58T, 58B, 59

Igna Spence/Tom Stack & Associates 137T

Tom and Therisa Stack/Tom Stack & Associates 96B, 97, 98T, 147R

Jean F. Stoick/Carl R. Sams II 109B

Diana L. Stratton/Tom Stack & Associates 24T, 130B

Patricia T. Sutton 65

Tom Till 32B, 61T, 61B, 67M, 83M, 111B, 119B, 125T, 133T, 151M, 179M

Larry Ulrich 77M, 117, 129B, 169M, 179B

Tom J. Ulrich 40, 122, 174T, 174B, 177T, 177B

Greg Vaughn/Tom Stack & Associates 186

Gary Vestal/Tom and Pat Leeson 48 (2nd from top)

Tom Walker 5B, 33T, 48 (2nd from bottom), 49T, 67B, 108B, 154B, 156T, 190, 193, 194B, 197, 202, 203T, 203B, back cover bottom

Michele Westmorland 94, 96T

Stuart Westmorland 99

Art Wolfe 8L, 27B, 31T, 44, 55T, 75B, 119T, 125B, 146T, 151T, 169B, 173, 180, 182, 184, 189B, 199M, 199B, 209B, 210/211

Norbert Wu 19, 42, 43T, 43B, 43BR, 90, 160, 162T

David Young/Tom Stack & Associates 93B

All maps by Karen Minot

Design and layout by Mary Kay Garttmeier

INDEX

Note: page numbers in italics refer to illustrations